T0275659

The Management of
# SCIENTIFIC INTEGRITY WITHIN ACADEMIC MEDICAL CENTERS

# The Management of
# SCIENTIFIC INTEGRITY WITHIN ACADEMIC MEDICAL CENTERS

**PETER J. SNYDER**
Lifespan Hospital System & Alpert Medical School
of Brown University
Providence, RI

**LINDA C. MAYES**
Yale University School of Medicine
New Haven, CT

**WILLIAM E. SMITH**
Chief Judge, United States District Court for the
District of Rhode Island
Providence, RI

Amsterdam • Boston • Heidelberg • London
New York • Oxford • Paris • San Diego
San Francisco • Singapore • Sydney • Tokyo
Academic Press is an imprint of Elsevier

Academic Press is an imprint of Elsevier
32 Jamestown Road, London NW1 7BY, UK
525 B Street, Suite 1800, San Diego, CA 92101-4495, USA
225 Wyman Street, Waltham, MA 02451, USA
The Boulevard, Langford Lane, Kidlington, Oxford OX5 1GB, UK

ISBN: 978-0-12-405198-0

**British Library Cataloguing-in-Publication Data**
A catalogue record for this book is available from the British Library

**Library of Congress Cataloging-in-Publication Data**
A catalog record for this book is available from the Library of Congress

For information on all Academic Press publications
visit our website at http://store.elsevier.com/

Typeset by TNQ Books and Journals
www.tnq.co.in

Printed and bound in the United States of America

Working together
to grow libraries in
developing countries

www.elsevier.com • www.bookaid.org

# DEDICATIONS

For my wise and loving mother, Susan J. Etkind, who has always been a wonderful supporter of my career and interests.

**Peter J. Snyder**

To Marion Mayes, my model for courage and resilience and Dr. Lawrence Cohen, my teacher and colleague in matters of research integrity.

**Linda C. Mayes**

For Christine, Katie and Allie who make it all worthwhile.

**William E. Smith**

# CONTENTS

# FOREWORD

As you thumb through the Table of Contents, the introductory chapters, or quickly skim through one of the detailed case studies presented in this book, you may be considering whether this volume—and its central topic—is of clear relevance to your professional practice as a scientist, clinician/scientist, academic medical center (AMC) administrator, graduate or postgraduate trainee, or scholar of modern research process. This book is unique in its presentation of issues, the competencies of its coauthors, and it is absolutely central to the culture, function, and reputation of any AMC, or medical research institute.

In my current role, I lead a large hospital Network, in one of the biggest population centers in North America. The Mount Sinai Health System and Network includes seven separate hospital campuses, a prestigious medical school (The Icahn School of Medicine of Mount Sinai), and multiple outpatient facilities. We credential over 6000 clinicians and scientists across 32 departments and 15 academic institutes. Research and innovation are core values and differentiating factors for our Health System and Network. Our commitment to these, as at other AMCs, is as central as our commitment to clinical excellence. At any one moment in time, there are many hundreds of professionals actively engaged in scientific research across virtually every facility and discipline within our system. All of this combined effort, and all of the scholarly productivity and successes of our faculty, would be rendered meaningless without constant vigilance over, and careful management of, research integrity. Our entire research enterprise, both as an individual medical system, but also as an entire academic society, hinges on our ability to protect and enhance the veracity of the scientific process.

We now live in an age in which research misconduct cases are regularly hitting the news and in which public trust in science and in fact in the overall quality and outcome of intervention and care within the medical profession is at an all-time low. Federal funding of science, is also greatly diminished. All of these critical environmental factors underscore the need to effectively manage ethical misconduct. Without a well-supported and active research integrity process and system in place, the pursuit of scientific discovery within an academic medical institution and indeed within our society is rendered meaningless. We need to clearly and openly look at these issues and commit to making our oversight and adjudication better.

In this volume, Professors Peter J. Snyder and Linda C. Mayes have formed a unique partnership, with Federal District Court Judge William E. Smith, to explore the application of NIH research integrity policy within hospital and AMC environments, and to better understand how the internal institutional management of misconduct affects—and is affected by—the United States' jurisprudence. The authors accomplish this large task through the presentation of a provocative introduction, an authoritative chapter on the legal and regulatory systems that guide the management of research conduct, the presentation of eight complex case studies, and by a summary chapter that provides "best practices" for those charged with such a large set of responsibilities. In choosing their anonymous cases for presentation, the authors have shown convincingly that, in most instances, there are rarely clear-cut "right" or "wrong" decisions to be made, but rather ambiguities and uncertainties that must be handled with care, consistency, wisdom, and high ethical standards; and where appropriate transparency and disclosure.

All three coauthors bring a wealth of experience, talent, good judgment, and knowledge to their work and to this book. Snyder, Mayes, and Smith have each enjoyed very productive and distinguished careers and the reader will most certainly benefit from their collective experience and advice. This volume will stimulate much discussion in the classroom (e.g., for graduate courses in science and ethics) and it will be a useful resource for both practicing scientists and for those who must serve their institutions by managing the oversight of research conduct.

Arthur A. Klein, M.D., President
The Mount Sinai Health Network
Executive Vice President—Mount Sinai Hospital
Executive Vice President—Icahn School of Medicine
New York City, NY, USA

# ABOUT THE AUTHORS

**Prof. Peter J. Snyder** has served as the Research Integrity Officer, as well as the Institutional Official, for the Lifespan Hospital System (Providence, RI) since 2008. He oversees the ethical conduct of research for approximately 350 investigators across a system of five hospitals that form the core teaching campuses for the Alpert Medical School of Brown University. Dr Snyder is the Chief Research Officer for this academic health system, and he also serves as a Professor of Neurology within the medical school. Dr Snyder has maintained an active research program for more than 20 years, and he has published widely in the fields of clinical neuropsychology, cognitive neuroscience, neuropharmacology and the history of the neurosciences. Dr Snyder is the Senior Associate Editor for *Alzheimer's & Dementia: The Journal of the Alzheimer's Association* (published by Elsevier).

**Prof. Linda C. Mayes** has served as the Special Assistant to the Dean, Yale School of Medicine (New Haven, CT) since 2008. In this role, she is also responsible for management of scientific integrity for the Yale School of Medicine. Dr Mayes is the Arnold Gesell Professor of Child Psychiatry, Pediatrics, and Psychology in the Yale Child Study Center, and her research integrates perspectives from child development, behavioral neuroscience, psychophysiology and neurobiology, developmental psychopathology, and neurobehavioral teratology. She has published widely in the developmental psychology, pediatrics, and child psychiatry literature. Dr Mayes is also trained as an adult and child psychoanalyst and she coordinates the Anna Freud Centre bridge program in social and developmental neuroscience and developmental psychopathology at the Yale Child Study Center. She is also a distinguished Visiting Professor in Psychology at Sewanee: The University of the South.

**The Hon. William E. Smith** is the Chief United States District Court Judge for the District of Rhode Island. He has served on that court for over 12 years. Judge Smith has developed an expertise and interest in the intersection between rapid advances in science and the work of the courts, in both the civil and criminal law. He has presided over a number of cases involving highly complex fields of scientific evidence. In addition, Judge Smith teaches various courses at the Roger Williams School of Law (Bristol, RI), including a course on the law of scientific and expert evidence; in addition, he serves on the board of The National Courts and Science Institute, an organization devoted to increasing the level of scientific knowledge and competency within the judiciary.

# ACKNOWLEDGMENTS

The authors would like to acknowledge the support, advice, expertise, and generosity of a number of individuals who helped make this published volume possible.

From the outset, a book on this controversial topic—including the presentation of detailed case studies—seemed to be a daunting task. Dr Snyder would first like to thank the General Counsel for the Lifespan Hospital System (Providence, RI), Kenneth Arnold, Esq., for his unfailing encouragement throughout this project. Ken has been an important mentor to Dr Snyder, and the organization and structure of this book (including our approach to the creation of anonymous case studies) resulted directly from Ken's questions and advice. Dr Snyder would also like to thank Therese Flynn-Eckford, Esq., who played the leading role in authoring the research misconduct policy provided in Appendix 2, for her advice and expertise. Dr Snyder would also like to thank Dr Tim Babineau, the President and Chief Executive Officer for the Lifespan Hospital System, for his consistent and unfailing support of the biomedical research mission of our institution (presently ranked 9th in the country, in terms of federal funding for research to private hospitals), and Ms Peggy McGill, Administrative Director for the Lifespan Office of Research Administration, for her wisdom and the often-needed patient advice of an "older sister!" Dr Snyder also thanks Dr Jack Elias, Dean of the Alpert Medical School of Brown University, for his dedication to advancing biomedical research across the entire Brown community, including its affiliated hospitals. Sophie Van Horne (Brown, 2014), a characteristically talented Brown University undergraduate student, assisted in compiling background material for this volume. Finally, Dr Snyder's spouse, Amy L. Snyder, M.D., and two children (Molly and Jacob), provided all the love, encouragement, and good humor to allow any writer to forge ahead with such a difficult project.

Dr Linda Mayes would first like to thank Dean Robert Alpern and Deputy Dean Carolyn Slayman of the Yale School of Medicine for the opportunity to work on matters of scientific integrity and conduct for the School of Medicine. In her role as Special Advisor to the Dean, Dr Mayes has become deeply appreciative of the complex social nature of research and the often conflicting motivations researchers are faced with. She is very grateful to her faculty colleagues who have voluntarily given hundreds of hours to carefully considering the many questions and allegations that come to a research integrity office. Indeed, her faculty colleagues continue to be exemplary teachers on how to thoughtfully and in a balanced matter hear the many sides of any scientific misconduct allegation. Dr Mayes also gratefully acknowledges her colleagues in the Yale General Counsel's office who provide support and guidance for Dr Mayes and the faculty committees reviewing the different allegations and concerns presented for consideration to Dr Mayes' office. Specifically, Robert Bienstock, Esq., senior associate general counsel and Harold Rose, Esq., senior associate general counsel, each in the Yale General Counsel's office have provided advice and review for a number of the issues and anonymous case descriptions in this volume and are both generous teachers and colleagues. Dr Mayes also gratefully acknowledges Marion Mayes, her mother, who consistently encouraged curiosity, fairness, and an appreciation for the richness of all human endeavors.

Chief Judge William Smith would like to first thank Dr Snyder for initially suggesting the collaboration that resulted in this book. It is because of his vision about the value of a work like this, and what a nonscientist might bring to the project, that Judge Smith became involved. Judge Smith is enormously grateful to his Research Assistant/Law Clerk, Michael Pabian, who took the laboring oar in the research for, drafting and editing of the Legal Landscape chapter and who contributed tremendous value to the rest of the work through his thoughtful comments and suggestions. Michael's acute legal mind, writing prowess, and incredible work ethic is second to none, and Judge Smith could not have done this without him. Michael was able to work on this project through the generous support of his law firm, Ropes and Gray, which allowed him to work for the judge for over a year as part of its New Alternatives Program. Thank you to Ropes and Gray for making him available. Judge Smith would also like to thank his intern, Scott Hefferman, a soon-to-be law student for his research and editing assistance as well. And finally, Judge Smith would like to thank Christine, Katie, and Allie Smith for all of their love and support in this and every project he has undertaken.

All three authors thank our editor at Elsevier B.V., Ms Halima Williams, for her active engagement as a partner in this effort—and for believing that this book will serve as an important resource for those who protect the veracity of the scientific process within academic medical centers.

## CHAPTER 1

# Introduction: Research Misconduct in Biomedical Research Institutions

*The management of research misconduct is central to the entire mission of an academic medical center. Without clear, transparent policies and procedures to guide the oversight of research ethics and scientific integrity, the veracity of the entire research enterprise is lost. The conduct of research is a human activity, and subject to the same human failings and errors of judgment that appear occasionally in any other field of human endeavor. In this chapter we describe the singular importance that the oversight of research integrity plays in the well-being of an academic medical center, and the we describe the rationale for this book and the chapters to follow.*

As we age, we look for simple lifestyle changes that might serve to protect our health and stave off dreaded diseases. And, at the same time, many adults enjoy a delicious glass of wine with a good meal. If we could pair together a plateful of cardiac health along with a glass of a full-bodied *Cabernet Sauvignon*? Fantastic! Over the past 20 years, there have been numerous reports concerning the benefits of moderate alcohol consumption (particularly red wine in first reports) in maintaining cardiac health (cf. Lavy et al., 1994; Bognar et al., 2013), decreasing the risk of certain forms of cancer (cf. Surh et al., 1999; Luo et al., 2013), and even as an incremental lifestyle protective factor for slowing the progression of age-related memory decline and Alzheimer's disease (cf. Russo et al., 2003; Wang et al., 2006; Pasinetti, 2012). The basic relationship(s) between modest alcohol ingestion and positive health effects have now been replicated by dozens of laboratories and clinical investigators, and so the risk of a false-positive association (a Type I error) between alcohol and/or other ingredients of alcoholic beverages (e.g., resveratrol), and certain health benefits, is unlikely. What remains unclear, however, is the strength of this relationship. That is, just how much benefit is derived from the glass or two of wine with dinner?

*The Management of Scientific Integrity within Academic Medical Centers*
http://dx.doi.org/10.1016/B978-0-12-405198-0.00001-1

One well-respected researcher to pose this question was Dipak K. Das, M.D., the former director of the cardiovascular research center at the University of Connecticut Health Sciences Center (Farmington, Connecticut). Dr Das had a reputation for scholarly contributions to this field, built over many years, with a specific focus on the health benefits of resveratrol, a naturally occurring phenol that is abundant in red wine. Dr Das had authored or coauthored more than 100 scholarly articles on this general topic, and his work had been highlighted in numerous popular magazines and television news programs (e.g., CBS network's *60 Minutes*). Although his general findings and conclusions have been replicated by other independent research groups, Dr Das's data seemed to suggest that red wine might be a powerful, key dietary approach to slowing the ravages of sedentary lifestyles and aging—even being referred to by the *Science Daily* website as "exercise in a bottle" (03 July, 2011: http://www.sciencedaily.com/releases/2011/06/110630131840.htm). Were his data too good to be true?

In early 2012, following a lengthy internal investigation by the university's Research Integrity Officer (RIO) and an appointed review committee, and in response to an "anonymous tip" in 2008, the institution found 145 separate instances of fabricated or falsified data by Dr Das, leading to the issuances of notification to 11 separate journals, in an attempt to correct the scientific record by asking these journals to publish 'retractions' of his papers. As a result of this investigation, the university declined and returned nearly $1,000,000.00 in Federal grant funding, closed his laboratory, and terminated his tenured professorship.

Another example was the work of psychologist Diederik Stapel, professor of cognitive social psychology and dean of the School of Social and Behavioral Sciences at Tilburg University in the Netherlands as well as founder of the Tilburg Institute for Behavioral Economics Research. Staple was regarded as one of Europe's top researchers with his findings about implicit contextual cues that motivate social behavior included findings such as how the presence of wine glasses changes meal behavior and table manners, that messy environments promote social discrimination, and that meat eaters are more antisocial than vegetarians. Widely published with over 250 peer-reviewed articles, book chapters, and conference proceedings, Professor Stapel was well-known and widely sought out as a collaborator given his innovative questions and ability to accomplish seemingly complex data collections in a timely manner. In 2011, an extensive investigation revealed that Professor Stapel had consistently fabricated data since at least 2004 with at least 55 published papers containing fraudulent data. In a pattern that appears to have extended well over a decade, Dr Stapel would team up with a student or colleague to design a study to test one of the collaborator's hypotheses. Dr Stapel would carry out the study and process the data. He provided his collaborator with a data file ready for analysis that was in reality entirely fabricated. In August 2011, three of Stapel's Ph.D. students brought forward allegations of data fabrication because they had been unable to obtain raw data from Dr Stapel when they repeatedly requested the files. Indeed, a number of his students graduated with their doctorates without ever completing an experiment. The investigation spread beyond Tilburg University to include the University of Amsterdam where Dr Stapel received his doctorate and the University of Groningen where he was employed prior to his professorship at the Tilburg University. The University of Amsterdam revoked his doctorate and Stapel resigned his professorship in the midst of ongoing investigations revealing more instances of complete data fabrication (Jarrett, 2013).

What motivated both of these scientists? What might compel an accomplished, well-trained scientist to risk his entire career in this manner? While there is excellent work that supports the health benefits of resveratrol as part of a healthy diet, it turns out that Dr Das's work—and, in fact, a good portion of his professional career—was proven to be disreputable. Did Dr Das simply believe that because he had proven his basic hypothesis to be correct early on, that all of his subsequent papers were less important and did not actually require valid data? Did Dr Stapel feel so confident of this observations and theories that he did not require data to confirm his beliefs? Indeed, when interviewed while the investigations were ongoing, Dr Stapel said: "I did not withstand the pressure to score, to publish, the pressure to get better in time. I wanted too much, too fast. In a system where there are few checks and balances, where people work alone, I took the wrong turn (Brabants Dagblad, October 2011, translated from Dutch)".

Still, neither Drs Das nor Stapel really address what motivated them to begin and continue data fabrication despite early successes and respect by the scientific community. To frame these questions more broadly, we might ask what are the motivating forces that would lead a scientist to commit acts of research misconduct (including, but not limited solely to fabrication of data), in order to advance his or her own research and career? Science is a social enterprise; and like all human activities, science is driven as much by the hopes, dreams, and vested interests of its practitioners as it is by the rational quest for discovery and understanding. Indeed, the wish and need for professional advancement and progression are strongly motivating, especially when science is conducted within the academy where scientific 'productivity' and reputation are the currency of faculty promotion. As scientific researchers, our behavior and professional conduct are molded by many hard years of education and training in scientific methodology. We grow comfortable with exploring the boundaries of current understanding. Yet unlike many professions that require their practitioners to limit their judgment and actions to factual record, scientists routinely glide along the razor's edge between the current state of factual knowledge and the wonder of new discovery. As such, the self-perceived ability to remain cautious in interpreting data, to recognize the limits of our own knowledge, and our deep appreciation for the awe-inspiring complexity of our natural world and the phenomena that we study, all become part of our professional identity.

As scientists, we are proud of our finely honed intellectual skills; and our beliefs in our own objectivity and integrity rest on our adherence to proper scientific methods and the testing of refutable, falsifiable hypotheses. And yet occasional scientific misconduct occurs. Perhaps it is simply because, in spite of our training, experience, and skills, we are human, and prone to the same moral and ethical failings that are characteristic of many segments of our larger society and culture. We want to advance our careers, to be promoted within our institutions or to be attracted by enticing career opportunities elsewhere, to provide needed resources to our students and fellows, to complete important work that will be remembered long after we are gone, and to provide financially for our immediate and extended families. All of these goals and ambitions are advanced by designing, accomplishing and publishing new research, building and managing valuable resources, attracting media attention, and garnering the support of our professional peers and home institutions. Our reputations, our authority, and ultimately our career and economic success depend on these key accomplishments and, broadly speaking, the pursuit of these goals often seems to be at the root of scientific misconduct when it occurs.

In a poignant essay, entitled "Conduct, Misconduct and the Structure of Science" Professors James Woodward and David Goodstein teach us that the most admirable of ethical principles—the very ideology that Western scientists are reared on—lie in direct conflict with the actual practice of science as an occupation and social endeavor (Woodward and Goodstein, 1996). These ethical tenets arise from the earliest theory of scientific method, as espoused by the English Renaissance philosopher and statesman, Sir Francis Bacon (1561–1626). Bacon, who was influenced in his thinking by Plato, asserted that scientists must act as disinterested observers of nature, with minds that are free of prejudices or preconceptions. With this belief held as central to competent practice as a scientist, Woodward and Goodstein offer a list of 15 corollary principles that conform to the Baconian ideal. Who among us would not generally agree with this sampling of principles to govern ethical conduct as a scientist?

- A scientist should never be motivated to do science for personal gain, advancement, or other rewards.
- When an experiment or an observation gives a result contrary to the prediction of a certain theory, all ethical scientists should question that theory.
- Scientists must report what they have done so fully that any other scientist can reproduce the experiment or calculation; science should be an open book, not comprised of guarded secrets.
- Financial support for scientific work and access to scientific facilities should be shared democratically, not concentrated in the hands of a favored few.

Woodward and Goodstein argue that although we aspire to live by these principles, and others also listed in their essay, they are "defective," they undermine the logical structure of science, and they are not consistent with the reality of scientific careers (especially in academic biomedical institutions). As noted above, scientists are motivated by the opportunity to make a lasting mark on their communities, be remembered by their peers for good work, provide well for their children and families, and achieve other markers of success as a professional, spouse, and care-provider. Further, scientists compete with their peers for limited funding support for their work, and within the academy, scientists are always motivated by the need for professional advancement as much if not more than the excitement of discovery and advancement of knowledge. These complex and often conflicting motivations often lead to temptations to publish too soon, to overinterpret a finding, to selectively publish data consistent with a theory or previous findings, to oversimplify a problem or solution, as well as more patently unethical subterfuge of the scientific process (e.g., stealing, falsifying or fabricating data, or intentional misrepresentation). At the same time, these very same complex motivating factors are not inherently corruptive and often work to propel scientific discoveries forward. Woodward and Goodstein note that "behavior that may seem at first glance morally unattractive, such as the aggressive pursuit of economic self-interest, can, in a properly functioning system, produce results that are generally beneficial."

There are good reasons to seriously consider the validity of the Woodward and Goodstein argument, that the corollary Baconian principles of scientific conduct are often in conflict with the complex personal, professional, and cultural motivations impacting the scientific enterprise, and this conflict shapes how science is both practiced in reality and also what brings vitality and innovation to the scientific process. As one example, the philosopher, Philip Kitcher, has made the point that because the first person who makes a scientific discovery usually attracts most of the credit, investigators are encouraged to pursue a broad range of different lines of inquiry—including avenues of research that may be thought by many to be associated with smaller chances of eventual success—maximizing their own chances of meeting success with uncovering an "unpolished gem" (Kitcher, 1990). This, in turn, leads to a wider diversity of concurrent lines of scientific inquiry, and as a result, it is more likely that an incorrect majority opinion in a given field might be overturned more rapidly. Woodward and Goodstein wisely caution us, in our design of "institutions and regulations to discourage scientific misconduct, that we not introduce changes that disrupt the beneficial effects that competition and a concern for credit and reputation bring with them" (Woodward and Goodstein, 1996). In other words, time honored as the Baconian principles may be, it is important that we do not build our regulatory structures solely around these corollaries.

If we accept the premise, which seems inevitable, that ambition and competition are necessary forces that propel scientific discovery forward, and that with this comes the temptation to cheat, then could the answer be to identify what types of individuals are most likely to engage in scientific misconduct, such as plagiarizing, falsifying, or fabricating data? While there is an obvious allure to this approach—it is scientific after all—there is little reason to believe that it is the solution. A detailed analysis of 146 separate misconduct cases, handled by the NIH Office of Research Integrity (ORI) between 1992 and 2003, reveals that roughly "one-third of the [individuals accused of misconduct] were support staff, one-third were post-doctoral fellows or graduate students, and one-third were faculty" (Kornfeld, 2012). The majority of these cases (66%) involved alleged falsification of data, and an additional 45% of cases also involved, or primarily included, the outright fabrication of data. A minority of these cases (12%) centered on accusations of plagiarism (Kornfeld, 2012). Although the author of this review attempted to categorize the types of individuals into various personality categories who might be most prone to engage in such behavior, such personality-typing is, at best, speculative at this point without formal evaluation. The fact is, with hundreds of thousands of scientists and students engaged in scientific research across North America alone, there will undoubtedly always be a small minority of individuals who are more prone than others to engage in misconduct. Thus, with proportionately so few individuals guilty of serious scientific misconduct, it is difficult to pinpoint specific personality traits or conditions that make such conduct more likely; but nonetheless there has been a regular drumbeat of high-profile ethical trespasses in science, throughout the history of modern research. One can look back easily to the celebrated unveiling of Piltdown Man by Charles Dawson in 1912, or the fraudulent series of case reports by Andrew Wakefield, in 1998, linking the vaccine for measles, mumps, and rubella as a cause for autism in children (Wakefield et al., 1998). There are many such easy examples to be found, and this latter example still exerts a chilling ripple-effect on public health and welfare—as scores of parents still refuse vaccines for their children without any credible data supporting their fears.

There is really no need to go back to the twentieth century to find examples of scientific misconduct that directly undermine public trust in scientific researchers and that tarnish the reputations of the institutions that support their work. One-and-a-half decades into the twenty-first century, we already have plenty of new cases to choose from, and ones that involve individuals whom we have considered to be at the very pinnacles of their careers—with everything to lose. Consider the case of Marc Houser, the celebrated psychologist and professor at Harvard University, who was found guilty on eight counts of scientific fraud by the NIH ORI in September of 2012. Dr Houser was an internationally renowned researcher and leader in the growing field of evolutionary psychology. His research was widely influential in that field, and he was a highly sought-after lecturer and author—by all metrics a veritable "rock star of science." In August 2010, Harvard University confirmed that he was found guilty of data fabrication by an internal review board, thus leading to a further investigation by the ORI. Specifically, the Harvard panel determined that Professor Houser intentionally produced "irregularities in data acquisition, data retention, and the reporting of research methodologies and results." Houser was subsequently barred from receiving Federal grant funding, from teaching in the Department of Psychology, and in 2011 he lost his tenured faculty position at the university.

Although some of his work has been retracted and/or publicly questioned (including by several of his own graduate students and postdoctoral fellows), the entire field has been left wondering what portions of Houser's prodigious body of work were "real" or valid, and what parts of his work should not be trusted. Houser's unethical behaviors also call into question the role and responsibilities of his coauthors on his many publications. One of his many coauthors stated that he "wrote the introduction and conclusions sections [of a research publication] but never reviewed the raw data, but just the data summaries." As we will discuss in two of our case studies in this volume (Chapters 7 and 8), there is just no simple answer to what the specific responsibilities of coauthors are, particularly when the research is highly complex and requires the collaboration of specialists from multiple disciplines—with no one individual in a position of competence to directly oversee every facet of a specific project. Although some have argued that each coauthor on a paper must assume full responsibility for the entire contents of the published work (cf. Carr, 2009), others contend that such a requirement would impede progress as scientific technologies, data informatics, and theory become more complex and cross the traditional boundaries of various disciplines. Further, many laboratories are large with teams of junior faculty, graduate students, and postdoctoral fellows working under, at least in name, the director of the lab. In highly productive labs such as these, it is very difficult for the lab director to oversee all details of the scientific process, data analysis, and manuscript preparation no matter how well intentioned and thorough the lab director may be.

Whatever motivates it, the reality is that ethical misconduct has been with us since the beginnings of human scientific exploration, and is arguably reaching "epidemic" proportions: the number of false claims made in published research, particularly in the biomedical sciences (Ioannidis, 2011), as indexed by a frightening rise in published retractions of articles in scientific journals, has grown dramatically in the past four decades (approximately a 10-fold increase in retracted publications, mostly due to fraud, since 1975; Fang et al., 2012). Our ability to effectively police and hopefully prevent or minimize the effects of scientific misconduct is essential—the credibility of our collective work as researchers rests entirely on this principle. Without codified, rigorous, and enforceable means to protect the veracity of research, the public support of our activities (amounting to nearly $140 billion for all types of US federally funded research support in the 2013 congressional budget) will erode and disappear.

Here is what is at stake: The rising tide of scientific misconduct calls into question public trust in science, at a point in time when US federal biomedical research funding is increasingly being curtailed. As a result, it is getting harder to retain the brightest students and to encourage them to both pursue careers in science and to remain confident that they can provide stable incomes to support their families. Adding to the stakes, these challenges facing talented young scientists are coming at a moment in history when we are tantalizingly close to major breakthroughs that could lead to improved treatments for some of our most devastating of human diseases (e.g., neurodegenerative diseases, neurodevelopmental disorders, various forms of brain cancers, osteoarthritis, autoimmune diseases, and psychiatric disorders such as depression and schizophrenia), and at a time when our legal system is increasingly called upon to resolve high stakes disputes that turn on scientific questions. We believe the stakes could not be higher and the need for rigorous and effective systems to investigate and manage allegations of scientific misconduct has never been greater.

The purpose of this book is to assist academic medical institutions to effectively maintain that first-line of defense against breeches of scientific integrity, as mandated within the United States by the NIH ORI. Two of us (Professors Peter J. Snyder and Linda C. Mayes) serve as RIOs for our home institutions[1], and we seek to explore both the strengths and weaknesses of the ORI policy on management of scientific integrity by presenting eight complex cases. For each of these cases, we intend to show how we applied both the ORI policy and our own institution's local interpretation of that policy, to guide our decision making. All of the cases we have chosen were resolved internally, without the initiation of a formal investigation by the NIH/ORI. Because so few cases are pursued by NIH/ORI, we believe this selection of cases reflects the kind of matters that occur with regularity in academic research institutions. In our experience, we rarely see scientific integrity cases that have simple findings of easily identified wrongful conduct. Rather, as we know from our experience as practicing scientists ourselves, actual "truth" is hard to identify, and it is far more often the case that we are analyzing cases made up of "shades of gray" and not "black-and-white" sets of facts. We have found, in each of these cases, that our ultimate decisions and actions, as the RIOs, deciding officials, and/or deans for our institutions, are arrived at after accounting for all sorts of nuanced influences—and none of this was "taught" to us when we accepted our current roles and responsibilities. There was no available guide or "casebook" to refer to that covers how to consider these nuances and shades of gray. We hope that this book will, to some degree, serve this purpose for others who are charged as RIOs, as well as to serve as a broader discussion of how

[1] The RIO, for the Yale University School of Medicine, holds the title of "Special Advisor to the Dean".

scientific conduct is managed amongst ethicists, academic medical administrators, graduate students and postdoctoral fellows, the legal profession, and all those interested in the interface between the practice of science and the social behavior of the scientists themselves.

In order to broaden the perspective, we have added an "outsider" to our group: Judge William Smith, a federal trial court judge with an interest in the intersection of law and science. In order to place the cases into the broader legal and regulatory context in which scientific research misconduct investigations take place, Judge Smith has provided an overview of the law surrounding this field. He researched the cases and worked with us to pose questions for discussion. In many ways, the conduct of a scientific integrity review within an academic medical center parallels actual legal investigations, even if not handled as a legal matter within the university or hospital system. We sequester evidence, we interview witnesses, we impanel committees to review all case materials, and we adjudicate claims and disputes. So it is useful to think about questions such as whether our activities potentially impact civil and criminal procedures that could be initiated as a result of misconduct claims being leveled against one or more faculty members. When is it appropriate for an investigator to contact legal counsel? When should the federal regulatory agency and/or outside law enforcement agencies be notified of potential misconduct or criminal activity? Are there steps that could be taken (or avoided) that would lessen the likelihood of a legal challenge to either the process or the outcome of the investigation? To this end, we have endeavored in each of our case studies to pose a variety of critical discussion questions about how the case was handled, whether mistakes were made or improvements could be suggested; these questions derive from our collective experience in research, medicine, and law.

Admittedly, we have carefully chosen a selection of some particularly challenging scientific conduct cases for study in this volume. It is entirely reasonable to expect that others, in similar positions and presented with the same information, would come to different conclusions for at least several of these vignettes. But our goal is not to prescribe a one-size-fits-all script for dealing with research misconduct investigations. We believe there may be various approaches and methods that will effectively advance the goals and requirements set forth in the US *Federal Policy on Research Misconduct* (Appendix 1). This serious and important activity is managed in slightly different ways by different institutions, including our own two medical centers in New Haven, Connecticut, and Providence, Rhode Island. These variations are not only acceptable, but probably necessary in as much as the policies and procedures of an institution must account for and function within the particular culture of the institution. As a reference guideline, we provide examples of how our institutions have responded to the federal mandate—the Lifespan Hospital System (Providence, Rhode Island) policy on academic misconduct is provided in Appendix 2, and the Yale University policy is provided in Appendix 3.

The intent of this volume is to assist those working in this field to understand in practical terms the scope and limits, the strengths and weaknesses of the statutory and regulatory system that guides the management of scientific misconduct, and to understand the intersection between these activities and the larger legal system. We hope that this approach will not only provide a useful guide to the process, but will also stimulate questions and encourage discussion in this critically important area.

## REFERENCES

Brabants Dagblad, 2011. Stapel Betuigt Openlijk 'Diepe Spijt' (Translated from Dutch). http://bd.nl/nieuws/tillburg-stad/stapel-betuigt-openlijk-diepe-spijt-1.121338.

Bognar, E., Sarszegi, Z., Szabo, A., Debreceni, B., Kalman, N., Tucsek, Z., Sumegi, B., Gallyas Jr, F., 2013. Antioxidant and anti-inflammatory effects in RAW264.7 macrophages of malvidin, a major red wine polyphenol. PLoS One 8 (6), e65355.

Carr, W., 2009. Prevailing truth: the interface between religion and science. In: Snyder, P.J., Mayes, L.C., Spencer, D.D. (Eds.), Science and the Media: Delgado's Brave Bulls and the Ethics of Scientific Disclosure. Elsevier, Amsterdam, pp. 107–121.

Fang, F.C., Steen, R.G., Casadevall, A., 2012. Misconduct accounts for the majority of retracted scientific publications. Proc Natl Acad Sci 109, 17028–17033.

Ioannidis, J.P.A., 2011. An epidemic of false claims. Sci Am Mag 304, 16.

Jarrett, C., 2013. Stapel—final report. Psychol (News) 26 (Part 2), 88.

Kitcher, P., 1990. The division of cognitive labor. J Philos 87, 5–22.

Kornfeld, D.S., 2012. Perspective: research misconduct: the search for a remedy. Acad Med 87 (7), 877–882.

Lavy, A., Fuhrman, B., Markel, A., Dankner, G., Ben-Amotz, A., Presser, D., Aviram, M., 1994. Effect of dietary supplementation of red or white wine on human blood chemistry, hematology and coagulation: favorable effect of red wine on plasma high-density lipoprotein. Ann Nutr Metab 38 (5), 287–294.

Luo, H., Yang, A., Schulte, B.A., Wargovich, M.J., Wang, G.Y., 2013. Resveratrol induces premature senescence in lung cancer cells via ROS-mediated DNA damage. PLoS One 8 (3), e60065.

Pasinetti, G.M., October 2012. Novel role of red wine-derived polyphenols in the prevention of Alzheimer's disease dementia and brain pathology: experimental approaches and clinical implications. Planta Med 78 (15), 1614–1619.

Russo, A., Palumbo, M., Aliano, C., Lempereur, L., Scoto, G., Renis, M., 2003. Red wine micronutrients as protective agents in Alzheimer-like induced insult. Life Sci 72 (21), 2369–2379.

Surh, Y.J., Hurh, Y.J., Kang, J.Y., Lee, E., Kong, G., Lee, S.J., 1999. Resveratrol, an antioxidant present in red wine, induces apoptosis in human promyelocytic leukemia (HL-60) cells. Cancer Lett 140 (1–2), 1–10.

Wakefield, A.J., Murch, S.J., Anthony, A., Linnell, J., Casson, D.M., Malik, M., Berelowitz, M., Dhillon, A.P., Thomson, M.A., Harvey, P., Valentine, A., Davies, S.E., Walker-Smith, J.A., 1998. Ileal-lymphoid-nodular hyperplasia, non-specific colitis, and pervasive developmental disorder in children. Lancet 351, 637–641.

Wang, J., Ho, L., Zhao, Z., Seror, I., Humala, N., Dickstein, D.L., Thiyagarajan, M., Percival, S.S., Talcott, S.T., Pasinetti, G.M., 2006. Moderate consumption of Cabernet Sauvignon attenuates Abeta neuropathology in a mouse model of Alzheimer's disease. FASEB J 20 (13), 2313–2320.

Woodward, J., Goodstein, D., 1996. Conduct, misconduct and the structure of science. Am Sci 84 (5), 479–490.

## CHAPTER 2

# Research Integrity: The Legal and Regulatory Landscape

## 1. INTRODUCTION

The integrity of institutional research is policed under a comprehensive set of self-effectuating regulations known as the Public Health Service Policies on Research Misconduct. These regulations create a multilayered approach to individual and institutional misconduct investigations in scientific research. The regulatory scheme is layered in the sense that it imposes first-line investigation and reporting obligations on the institution in which the alleged misconduct occurs with secondary proceedings occurring, if at all, within the regulatory agency, the United States Department of Health and Human Services (HHS). HHS has delegated the lion's share of its responsibility for addressing research misconduct to the Office of Research Integrity (ORI),[1] created by Congress in the National Institutes of Health Revitalization Act of 1993.[2]

This chapter will review in some detail how the regulations work: the obligations they impose and the procedures they employ. Following a general overview, and a sketch of the institutional/ORI process, we will examine various kinds of legal challenges that have been brought against the regulatory scheme itself, as well as specific challenges to actions taken by institutions pursuant to the regulations, and some of the lessons these cases teach; finally, we will attempt to summarize the legal mosaic that this combination of statutory, regulatory, and common law (or case-law) has created.

## 2. THE INSTITUTIONAL AND AGENCY PROCESS

We begin with the regulatory scheme created by Congress and HHS. HHS regulations[3] place the initial responsibility for responding to research misconduct with "institutions,"[4] which include "colleges and universities" as well as certain "research laboratories" and "research and development centers."[5] Institutions that apply for or receive support from the Public Health

Service (PHS), a unit within HHS,[6] are required to, among other things, put in place "written policies and procedures for addressing research misconduct."[7] These policies must include protection of the parties' confidentiality, protocols for handling the research record and evidence, institutional actions in response to findings of research misconduct, efforts to protect the reputations of individuals involved in the proceedings, and full cooperation with ORI.[8]

The regulations define "research misconduct" as "fabrication, falsification, or plagiarism in proposing, performing, or reviewing research, or in reporting research results."[9] The definition "does not include honest error or differences of opinion,"[10] and it requires that the misconduct be "committed intentionally, knowingly, or recklessly."[11] Research misconduct must also involve "a significant departure from accepted practices of the relevant research community."[12]

It is important to note, however, that institutions are free to adopt standards for research misconduct broader than those prescribed by the regulations,[13] and in fact, some institutions have done this. Both Yale University and the University of Medicine and Dentistry of New Jersey (now known as Rutgers Biomedical and Health Sciences), as two examples, arguably exceeded the scope of federal regulations by prohibiting "duplicative publication," or redundant publication of research that has already been reported in a published article.[14] Similarly, the University of Maryland, Baltimore, defines misconduct to include "[i]mproper experimental manipulation," "[i]mproper assignment of credit," "[a]buse of confidentiality," and "[m]isappropriation of funds or resources."[15] As of September 2000, an ORI study of 156 institutional policies found that slightly more than half those policies contained a definition of research misconduct broader than that used by the agency.[16]

When a Research Integrity Officer (RIO) receives an allegation of research misconduct, he or she must conduct a preliminary assessment of the allegation to determine whether an inquiry is warranted.[17] In making this determination, the RIO should consider the credibility and specificity of the allegation, whether it involves PHS support, and whether it falls within the definition of research misconduct.[18] There is no particular formula for assessing threshold credibility, but, as a practical matter, an RIO should keep in mind that he/she is not a rubber stamp for allegations of misconduct; the RIO should make a serious assessment as to whether the allegation seems well-grounded or merely speculative or vindictive. Moreover, the threshold burden is not a high one—but, in practical terms, must pass the "smell test." In other words, the allegation needs to appear credible on its face, to make sense and stand up to a common-sense assessment of credibility. Some institutions (e.g., Yale University) have in place a procedural step to help RIO's make this determination. For example, an initial step may be to seek consultation of members of the institutional staff or faculty with sufficient expertise in the science to review the allegations and/or materials as received and render an opinion to a hypothetical question: Were these allegations true, would they constitute academic misconduct? While the threshold for moving forward is not high with such a hypothetical, such a consultative step necessarily insures some consideration of the merit and substance of the allegation and provides the RIO with an initial consideration of the possible questions to guide a formally convened review committee.

If the RIO's preliminary assessment (with or without consultation from others) finds the allegation is credible, he or she must initiate an "inquiry," which involves "an initial review of the evidence."[19] Typically, the inquiry will be conducted by a committee appointed by the RIO.[20] At or before the beginning of an inquiry, the RIO must make a good faith effort to notify the accused (called the "respondent" in the regulations) and take all reasonable and practical steps to obtain custody of research records and evidence.[21] (These records, as well as additional materials generated during the institutional proceedings, generally must be retained for 7 years.[22]) After interviewing key witnesses and examining relevant evidence, the committee, in consultation with the RIO, will decide whether an investigation is warranted.[23] The inquiry must culminate in a written report,[24] and it is highly recommended that institutional counsel review the report for legal sufficiency.[25]

The regulations require that the respondent be notified of the inquiry's outcome[26] and be given an opportunity to review and comment on the inquiry report.[27] If the conclusion is that the allegation has merit, then an "investigation" is called for. The institution should conduct an investigation if it appears that the allegation falls within the definition of research misconduct, involves PHS support, and may have substance.[28] If the institution determines that an investigation is warranted, then notice to ORI is required.[29]

Once an internal proceeding has moved to the investigation stage, the institution must examine all research records and relevant evidence, conduct interviews, and pursue all leads.[30] The investigation, much like the inquiry that preceded it, will generally be conducted by a committee appointed by the RIO.[31] At or before the beginning of an investigation, the RIO must provide notice to the respondent and ORI, as well as take all reasonable steps to obtain custody of research records and evidence that were not previously sequestered during the inquiry.[32] At the outset, the RIO should meet with the committee to review the charge as well as the applicable procedures and standards.[33] On (somewhat rare) occasions the accused may seek representation by counsel during the investigation. This, of course, is his right. If counsel is engaged, he or she should be afforded all the courtesies one would expect, but should not be allowed to obstruct the process. The respondent is expected to cooperate in the investigation, and if he does not cooperate an adverse inference may be appropriate.

After completing its investigation, the institution must provide the respondent with a draft of the investigation report and allow him or her an opportunity to comment on it.[34] Before submitting the report to the respondent, institutional counsel should review the draft report for legal sufficiency.[35] Once comments are received and considered, the investigation culminates in a final report,[36] which must be provided to ORI.[37]

Generally, institutions should carry inquiries and investigations to completion. However, if, at any stage in the proceedings, the institution plans to close a case because the respondent has admitted guilt or entered into a settlement, the institution must notify ORI of this development.[38] ORI may, after conducting an oversight review, approve closure of the case or direct the institution to complete its process.[39]

The self-examination and self-reporting requirements discussed above are without question a heavy obligation: failure to properly respond to allegations of misconduct may result in severe sanctions for the institution, including debarment from receiving government research funds.[40] Moreover, additional adverse consequences are possible including civil liability under the False Claims Act (FCA). Given the magnitude of what is at stake (continued federal research dollars), the incentives should be sufficient for research institutions to adopt the required protocols and treat all allegations of misconduct seriously.

Federal regulations empower ORI to "respond directly" to an allegation of research misconduct "at any time before, during, or after" institutional proceedings.[41] ORI's options include conducting an independent assessment of the allegation, forwarding the allegation to the institution or appropriate HHS department for inquiry or investigation, and reviewing the institution's findings and process.[42] In reviewing institutional research misconduct proceedings, ORI may consider materials generated at the institutional level and/or "[o]btain additional information."[43] Ultimately, ORI is empowered to "[m]ake appropriate research misconduct findings and propose HHS administrative actions."[44] Included among the possible administrative actions that HHS may take are debarment and suspension,[45] meaning "the Government wide exclusion…of a person from eligibility for Federal grants, contracts, and cooperative agreements" under the relevant regulations.[46] Other less severe sanctions include correction of the research record, letters of reprimand, imposition of certification requirements, termination of a grant or contract, restriction on activities under an active grant or contract, special review of requests for PHS funding, imposition of supervision requirements, and prohibition on participating in any advisory capacity to PHS.[47]

Generally, when ORI makes a finding of research misconduct it notifies the respondent in a document called a "charge letter."[48] A charge letter may be thought of as a kind of "indictment," but in the civil (as opposed to criminal) law arena. The charge letter should set forth the facts supporting the misconduct charge in sufficient detail so that the accused may understand the basis for the allegation. A respondent may contest ORI research misconduct findings and administrative actions by requesting a hearing before an Administrative Law Judge.[49] If, however, the respondent fails to contest the charge letter within 30 days, ORI's findings become the final HHS action.[50] The one exception to this rule is that an HHS "debarring official" must make the final decision on all debarment or suspension actions.[51]

It is important to understand that HHS performs a dual role here, common to many federal agencies, of investigation and adjudication. The Administrative Procedure Act (APA) is the governing statute that sets out guidelines for all federal agencies to follow in prescribing general rules[52] and adjudicating the rights of specific parties.[53] The APA also provides standards for judicial review of agency actions.[54]

In the specific context of research misconduct matters, ORI performs the investigative role (and acts as the prosecutor if the case proceeds to adjudication); the quasi-judicial role is performed by a cadre of judges within the HHS Departmental Appeals Board (DAB) known as Administrative Law Judges (ALJs). ALJs are not part of the judicial branch of government; rather they are employees of the individual federal agency which generates the cases they hear. Nevertheless, ALJs enjoy many of the attributes of an independent judiciary, and function much like federal judges when adjudicating enforcement proceedings pursuant to the governing regulations of the agency within which they work.

When ORI issues a charge letter, the accused individual must respond by either accepting or challenging the findings of research misconduct and proposed administrative actions. If the respondent contests the charges, the matter will be scheduled for a hearing before an ALJ in much the same way a case is scheduled for trial in state or federal court. The parties to an ALJ hearing, the respondent and ORI, have various procedural rights, which are outlined in the regulations. Specifically, the parties have the right to be represented by an attorney, participate in conferences, conduct discovery, agree to stipulations of fact or law, file motions, present evidence, cross-examine witnesses, present oral arguments, submit post-hearing briefs, and submit materials under seal where necessary.[55] However, just like in federal and state courts, not all matters require a hearing. The ALJ may decide cases by a process called "summary judgment" "where there is no disputed issue of material fact", and the issue involves only applying the law to those facts.[56]

ORI is required to prove findings of research misconduct and the reasonableness of proposed administrative actions by a preponderance of the evidence.[57] In other words, it must prove "that the fact at issue is more probably true than not."[58] This is the burden of proof used commonly in civil proceedings in state and federal court. Importantly, the burden of proof is *not* akin to that used in criminal matters (proof beyond a reasonable doubt) or even that used in some higher burden civil cases (clear and convincing evidence), such as civil commitment proceedings or cases terminating parental rights. A preponderance of the evidence is, from a legal point of view, a relatively low threshold of proof, especially given the dramatic potential consequences to the charged parties.

After the parties have presented the case at a trial and made their final submission in post-hearing briefs, the ALJ must issue a written ruling, setting forth proposed findings of fact and conclusions of law.[59] This ruling constitutes a recommended decision to the Assistant Secretary for Health, who may modify or reject the ruling if he or she finds it to be "arbitrary and capricious or clearly erroneous."[60] If the Assistant Secretary does not notify the parties of its intention to review the ALJ's decision within 30 days, that decision becomes final.[61]

The regulations allow HHS to settle a research misconduct proceeding "at any time."[62] And in fact, most ORI cases resulting in a finding of misconduct are closed by means of "a voluntary settlement agreement with the respondent."[63]

# 3. OTHER LEGAL CONSEQUENCES FOR RESEARCHERS ENGAGING IN RESEARCH MISCONDUCT

As discussed above, individuals who commit research misconduct face the prospect of severe administrative sanctions, including debarment or suspension. However, these potential agency sanctions represent merely the tip of the iceberg with respect to such individuals' potential legal exposure. Research misconduct may lead to crushing civil liability or even imprisonment.

## 3.1 False Claims Act

The FCA imposes civil liability on any person who "knowingly presents, or causes to be presented, a false or fraudulent claim for payment or approval."[64] The statute's reach also extends to an individual who "knowingly makes, uses, or causes to be made or used, a false record or statement material to a false or fraudulent claim."[65] The term "claim" includes "any request or demand...for money or property" that "is presented to an officer, employee, or agent of the United States."[66] Interestingly, private persons are permitted to bring FCA actions in the government's name.[67] FCA violators are generally liable for more than three times the damages sustained by the government.[68]

In the research misconduct context, plaintiffs have brought FCA suits against purported violators predicated upon the submission of allegedly false applications for government grants.[69] As a preliminary matter, the FCA does not require that the grant application itself contain false data;[70] this is because, as the statute makes clear on its face, the mere *use* of a false record is unlawful.[71] Thus, a defendant may be liable for submitting a grant application that relies on false data.[72]

Falsity is, however, a prerequisite to FCA liability, and it has proved a significant obstacle for plaintiffs in research misconduct cases. One example is the case of Kathryn Milam, who filed an FCA suit against the University of California and various university employees, alleging that they had submitted false data in connection with federal grant applications.[73] Milam, who worked at the University of California at San Francisco, was asked to replicate a study performed by one of the individual defendants.[74] The study involved treatment of brain tumor cells with a particular drug (BCNU), after the cells had been previously exposed to another drug (DFMO) or X-rays.[75] In attempting to replicate the study, Milam added a "rinse step," meaning she rinsed the "DFMO-treated medium" off the cells before adding BCNU.[76] Milam added this step because she believed the acidity of the medium was affecting her results.[77] After adding the rinse step, Milam was unable to replicate the study.[78]

Milam's FCA claims were predicated on the inclusion of the defendant researcher's findings, which Milam believed to be untrue, in federal research grant applications. The courts ultimately rejected this claim, characterizing the situation as "a legitimate scientific dispute," holding that "[d]isagreements over scientific methodology" are not sufficient to create FCA liability.[79]

*Milam* should not be read to imply that a researcher will invariably be able to escape FCA liability by appealing to scientific judgment or disagreement. In *Jones v. Brigham & Women's Hospital*, the court rejected just such an argument. There, the defendant researcher made "revisions" to his data, purportedly applying the same protocol that he used in his initial measurements. The court held that "scientific judgment" was not necessarily at issue and determined that a reasonable jury could find that the data were falsified.[80]

Another federal court has suggested that, at least in some circumstances, certification of compliance with applicable regulations may constitute a false claim sufficient to create FCA liability. Joan Luckey brought an FCA claim alleging, among other things, that her former employer, Baxter Healthcare Corporation, misrepresented compliance with federal regulations.[81] The court suggested that regulatory violations may serve as the basis for FCA liability where compliance with the applicable regulations is "a material condition to receiving payment from the government."[82] Baxter, a producer of plasma products, included with its shipments to government customers certifications of conformance with certain contractual terms. Among those terms was a requirement that Baxter comply with applicable statutes and regulations.[83] The court found that this certification "could constitute a claim";[84] however, it rejected Luckey's argument that Baxter's alleged insufficient testing of samples rendered its certifications false and, accordingly, granted Baxter's motion for summary judgment. Citing *Milam*, the court concluded that Luckey presented "insufficient evidence to demonstrate that her dispute with Baxter's whole blood testing process was anything other than a matter of scientific judgment."[85] It also pointed to the lack of evidence that the applicable regulations required the type of testing advocated by the plaintiff.[86]

Despite the court's ultimate ruling for the defendant, *Luckey* suggests that a plaintiff could bring a successful FCA claim predicated upon a defendant's failure to comply with federal regulations concerning research misconduct. In fact, one federal court was presented with an argument that the defendant institution violated the FCA because its inquiry into allegations of research misconduct violated HHS regulations.[87] In this case, the defendant had signed an "Applicant Organization" certification promising to comply with PHS terms and conditions.[88] While the appeals court ultimately affirmed the lower court's holding that the plaintiff had waived this certification claim,[89] it implied that a properly preserved claim could be successful. Indeed, federal regulations impose various requirements on covered institutions concerning how those institutions respond to allegations of research misconduct. Moreover, they require institutions seeking PHS funds to file an assurance, certifying regulatory compliance.[90] If an institution certifies compliance but ultimately fails to satisfy the regulations, a court may well find that it has submitted a false claim to the government.

In addition to proving that the defendant submitted a false claim, an FCA plaintiff must demonstrate that the defendant acted "knowingly."[91] This element of an FCA suit is not satisfied where a claim is "submitted as the product of the defendant's good faith professional opinion or judgment."[92] Accordingly, a federal court has found that the mere fact that a defendant researcher was on notice of and disregarded the plaintiff's concerns about data used in a grant application is not sufficient to establish knowledge of falsity.[93]

Even a plaintiff who proves that the defendant knowingly made a false claim may not be successful in an FCA suit. Such a plaintiff must also demonstrate that the falsity at issue was "material" to the claim.[94] The statute defines "material" to mean "having a natural tendency to influence, or be capable of influencing, the payment or receipt of money or property."[95] In the context of applications for government grants, materiality requires that the alleged falsity "have a natural tendency to influence the Application reviewers."[96] This element of an FCA claim seems easily satisfied where a grant application relies on falsified data. By contrast, materiality would not be so easily established where, for example, the alleged false statements involve minimizing the contributions of a graduate student whose efforts were, in any case, "not central" to the project.[97]

An interesting problem may arise where a plaintiff brings an FCA suit predicated upon allegations of research misconduct after ORI has declined to find that the defendant committed research misconduct. The court in *Milam* held that ORI's determination that the evidence of research misconduct was insufficient to warrant an investigation did not preclude the plaintiff's FCA claim.[98] The court explained that, in the case at hand, ORI did not conduct a hearing, and therefore the parties did not have an opportunity to litigate the issue.[99] The court went on to note that federal regulations required a higher level of intent for an ORI finding of research misconduct than that required under the FCA.[100] The vitality of this aspect of *Milam* is not clear for two reasons. First, the court's stated rationale for refusing to give ORI's decision preclusive effect is arguably limited to those situations in which the agency does not conduct a hearing. Moreover, the court relied on a prior version of the applicable regulations that was "silent on the level of intent required, except to exclude honest error."[101] The regulations now expressly define research misconduct to include reckless behavior.[102] In any case, while the *Milam* court refused to give the ORI report preclusive effect, it never-the-less held that the report was admissible under the Federal Rules of Evidence.[103] As a practical matter, this evidence is likely to influence the outcome at trial.

FCA plaintiffs are protected from retaliation under 31 U.S.C. Section 3730(h). That provision imposes liability on any defendant that discriminates against an employee "because of lawful acts done by the employee. . . in furtherance of" an FCA suit.[104] A plaintiff can bring a successful Section 3730(h) retaliation claim even if no underlying FCA suit is filed.[105] Courts have identified three elements of an FCA retaliation claim: "(1) the employee must be engaged in conduct protected by the statute, (2) the employer must know the employee was engaging in such protected conduct, and (3) the employer must have discriminated against the employee because of this protected conduct."[106] In *Luckey*, the federal district court held that the first element is not satisfied where the plaintiff's behavior is "directed at convincing [the defendant] to upgrade its procedures."[107] Rather, the plaintiff must present evidence that his or her actions "were related to exposing a fraud upon the government."[108] Relatedly, to satisfy the second element, a plaintiff must show that his or her conduct was sufficient to put the defendant on notice "not only of the voiced concerns and investigations of the employee, but that the employee's actions are related to the employer's alleged false claims to the government."[109] Finally, even if a plaintiff can satisfy the first two elements, an FCA retaliation claim requires proof of causation. The plaintiff's mere belief that the defendant's actions were motivated by his or her involvement in protected activities is not sufficient to satisfy this requirement.[110]

Section 3730(h) has been amended since *Luckey* was decided. That provision now protects employees against discrimination because of acts done in furtherance of an FCA suit "or other efforts to stop [one] or more violations" of the FCA.[111] This language is potentially broad enough to encompass conduct like that of the plaintiff in *Luckey*. While the plaintiff's actions in that case were not directed at exposing a fraud on the government, they were arguably designed to stop the alleged FCA violations by convincing the defendant to upgrade its procedures.

Another limitation on FCA retaliation claims is that Section 3730(h) applies only to actions in furtherance "of a *viable* FCA case."[112] Thus, defendants are entitled to judgment in their favor where the application containing the allegedly false statements was not a request for money and, for that reason, could not constitute a claim.[113]

## 3.2 Criminal Charges

On top of administrative actions and civil liability, research misconduct may result in criminal sanctions. *United States v. Keplinger* stands as a vivid example of the grave risks that falsification of scientific data entails. That case involved two studies conducted by Industrial Bio-Test Laboratories, Inc. (IBT) using rats to test the toxicity of particular substances.[114] Both studies resulted in the submission of reports to the federal government.[115] Ultimately, several defendants associated with IBT were convicted of mail fraud,[116] wire fraud,[117] and making false statements to the government.[118] The Seventh Circuit Court of Appeals affirmed the convictions, rejecting the defendants' numerous arguments. A review of the court's resolution of some of these issues demonstrates the potential breadth of the relevant statutes.

With respect to the first of two toxicity studies, the government alleged that the defendants underreported the mortality rate of rats exposed to the substance being tested. In support of this contention, the government relied on the fact that three documents generated by IBT contained three different sets of mortality data.[119] Moreover, all three mortality rates were lower than the one shown in internal IBT documents.[120] Lab technicians also testified that they observed deaths which were not accurately recorded.[121] The court found these facts sufficient to support a finding that the mortality data were false.[122] Contrary to the defendants' contentions, the government was not required to prove the true mortality rate. The court explained, "The very nature of the allegations suggests the inherent impossibility of showing the 'true' rate of mortality, since the accurate underlying data was never recorded."[123]

The government separately argued that the defendants fraudulently omitted the conclusions of Dr. Ribelin from their report to the government. Dr. Ribelin, an independent consultant, examined tissue samples from the rats used in the study and discovered degeneration in the epididymis, which is the excretory duct of the testes.[124] The defendants argued that they were not required to disclose Ribelin's conclusion because the testes, not the epididymis, were the target organs of the study.[125] The court rejected this contention, finding that "the jury was entitled to accept the government's alternative view that if treatment-related effects were found in any part of the rat, a safe level of dosage had not been established."[126] The court similarly disposed of the defendants' argument that an omission could not be used to support a mail fraud conviction. It explained that "omissions or concealment of material information can constitute fraud…cognizable under the mail fraud statute, without proof of a duty to disclose the information pursuant to a specific statute or regulation."[127] The court did, however, caution, "we do not imply every omission from a scientific report of potentially important information could form the basis of a mail fraud conviction."[128]

The Seventh Circuit Court of Appeals also affirmed the convictions arising from the second IBT study. The government alleged there that the defendants never actually collected certain data that they reported. In support of this claim, it presented evidence that the relevant tests were canceled and never rescheduled.[129] Moreover, the raw data from these tests could not be located.[130] The court held that "[p]roof of absence of records that would ordinarily exist if a particular event had occurred is properly admitted to show that the event did not occur."[131] With respect to the required mental state, the court concluded that "[a]lthough failure to investigate" missing data, "considered in isolation, may not be sufficient to demonstrate" a defendant's knowledge of falsifications, there was sufficient evidence for the jury to infer knowledge in the case at hand.[132] The court explained that the failure to investigate was presented "in combination with other evidence."[133]

Other possible criminal consequences arise where an individual provides false information to government officials investigating allegations of research misconduct. Such an individual may be charged with obstruction of justice under 18 U.S.C. Section 1505.[134]

## 4. PROCESS-BASED CHALLENGES

The inquiry and investigation process of the institution and the formal process of ORI outlined above create a thorough and presumptively fair process for reporting, examining, and prosecuting research misconduct.

Nonetheless, the procedures have been subject to numerous and varied legal challenges over the years. Most of these challenges have been unsuccessful; examples from reported cases nevertheless teach a number of important practical lessons from the real world about the handling of these research misconduct matters, where things can go wrong, and what can be done to mitigate the potential for problems.

## 4.1 Constitutional Due Process

We begin with the unfortunate story of Frances Edward Shovlin. After a series of conflicts with university administration,[135] Shovlin, a professor at the University of Medicine and Dentistry of New Jersey (UMDNJ), was named in an institutional investigation into alleged research misconduct.[136] The investigatory panel concluded that Shovlin was "only very peripherally involved" in the misconduct and recommended no sanctions against him.[137] Nonetheless, Paul Larson, the Senior Vice President for Academic Affairs, sent a memorandum to the Dean, listing Shovlin among the individuals against whom corrective action would be imposed.[138] That same day, Shovlin received two letters, copies of which were forwarded to other university officials and a representative of ORI, notifying him that his work would be subject to review for a period of 3 years.[139] Subsequently, the Board of Trustees decided not to grant Shovlin appointment as a Professor Emeritus.[140] Additionally, Shovlin was notified that his at-will adjunct faculty appointment had not been renewed.[141]

Shovlin responded to this barrage by bringing suit in federal court, alleging that Larson's actions violated his constitutional right to due process. It is important to note at the outset that Shovlin was only able to maintain a claim of a constitutional due process violation because UMDNJ was a public institution. The Due Process Clauses of the federal constitution (and states' constitutions) apply only to conduct by the government, not private parties;[142] thus, with limited exceptions, an aggrieved party may not bring a due process suit against a private college or university.[143]

The federal trial court rejected Shovlin's claim on the merits. To establish a procedural due process violation under the Fifth or Fourteenth Amendment, a plaintiff must demonstrate a protected "property" or "liberty" interest, which a government actor (here, the state university) has taken away.[144] Shovlin's "unilateral expectation" of appointment as Professor Emeritus was insufficient, as was his similar expectation that his adjunct appointment would be renewed.[145] The court also held that alleged damage to the plaintiff's reputation was insufficient to support his due process claim absent evidence of some "accompanying deprivation," such as loss of employment.[146]

Shovlin's case was doomed by lack of a concrete property interest—he was an at-will adjunct and had not yet received the "Emeritus" designation; a tenured professor at a public institution, on the other hand, may have a "constitutionally protected property interest" in his or her continued employment.[147] Similarly, a student accused of research misconduct may have a property interest in "an expected degree from a public institution."[148] So, the viability of any due process challenge will depend at the outset on whether the institution is a public actor and on the strength of the property claim.

Even if a respondent, unlike Shovlin, is able to successfully establish a cognizable property interest, he or she may not be entitled to a hearing before action is taken.[149] To determine whether due process requires a hearing *before* the deprivation of a protected interest, courts balance three factors: (1) the strength of the private interest at stake; (2) the risk of erroneous deprivation of that interest; and (3) the government's interest.[150] In the relevant contexts of education and employment, "[c]ourts have held that [ ] *post-deprivation* procedures, [such as] providing for a hearing to contest a challenged employment decision, are sufficient to satisfy due process."[151]

Additionally, if the institutional process ultimately results in a finding of no research misconduct, the respondent will be unable to demonstrate any deprivation of a protected interest and, thus, will not be successful on a due process claim.[152]

## 4.2 Breach of Contract

Accused researchers have also brought breach of contract suits challenging the process afforded in institutional research misconduct proceedings. Dr. Douglas Kerr brought one such suit against his former employer, Johns Hopkins University, alleging that the university failed to adhere to its own procedures in adjudicating his research misconduct case.[153] The court dismissed Kerr's claim, reasoning that several of the relevant policy provisions were not sufficiently "definite" to be enforceable.[154] Among the statements the court found to be merely "aspirational" were "the policies of conducting a 'sufficiently flexible' investigation, handling allegations of misconduct 'as confidentially as possible,' and collecting evidence in an 'objective, independent, unbiased, and thorough [manner].'"[155]

Another court has found that policies that are not "mandatory" or "exclusive" are also insufficient to form the basis of a breach of contract suit.[156] Thus, where more than one disciplinary procedure is available, the mere fact that the university chooses a procedure other than the one preferred by the plaintiff does not subject it to liability.

## 4.3 Suits by Victims

Individuals accused of research misconduct are not the only people who have challenged the sufficiency of institutional procedures for investigating alleged or potential misconduct. Alleged victims of research misconduct have brought similar process-based challenges. Ruey-Jen Hwu Sadwick, for example, a tenured professor at the University of Utah, sued the university and several of its employees for their allegedly deficient handling of Sadwick's allegations against her fellow faculty member, George Gray. Sadwick told university officials that Gray had sought to publish her research results as his own.[157] An institutional committee dismissed Sadwick's claims without a hearing.[158] Sadwick brought suit in federal court, alleging, among other things, that the university as well as several of its faculty and administrators appropriated her work without affording her due process.[159] The court dismissed Sadwick's constitutional claims against the three defendants alleged to have failed to conduct a proper investigation. In doing so, the court drew a distinction between "[d]irect misappropriation of trade secrets" and "failure to hold a hearing on the matter of whether a colleague has misappropriated trade secrets."[160] While clearly established law prohibited the former, it did not prohibit the latter.[161]

The *Sadwick* case further highlights the difficult position of plaintiffs attempting to sue a state university and its employees. To hold the individual defendants liable, it was not enough for Sadwick to prove that they had arguably violated her constitutional rights. This is because, as state employees, those defendants were protected by a doctrine known as "qualified immunity" and could only be held liable if their conduct violated "*clearly established* federal rights."[162] Similarly, the Eleventh Amendment, with a few exceptions,[163] will bar suit against the university itself as well as claims against university employees in their official capacities.[164]

## 4.4 Challenges to Agency Process

As described above, federal regulations allow ORI to review institutional investigations of alleged research misconduct and propose administrative sanctions. In addition to challenging the sufficiency of the institutional process, accused individuals have alleged that the federal regulations themselves provide insufficient process at the agency level.

Scott Brodie was barred from participating in federally funded projects for 7 years after the government determined that he had committed research misconduct.[165] Brodie brought suit in federal court challenging his debarment and arguing, among other things, that the ALJ's "summary disposition" of the case without a hearing violated due process.[166] The court rejected this argument, comparing Brodie's situation to that of a party in a civil case who loses on summary judgment.[167] It also held that application of the preponderance of the evidence standard in debarment proceedings did not offend the Constitution.[168]

A respondent in research misconduct proceedings must await a final agency action, such as the imposition of sanctions, before filing suit to challenge the sufficiency of the agency process.[169] Moreover, the result reached by the agency may moot any due process claim the respondent might have. The case of Dr Mikulas Popovic demonstrates the point. Dr Popovic was found to have committed research misconduct by ORI.[170] He appealed ORI's findings and, after a hearing, was ultimately absolved of any wrongdoing.[171] Nonetheless, Popovic brought suit, alleging that ORI violated his rights in various ways. More specifically, Popovic argued that

> investigators failed to give him timely notice of the nature of the proceedings and charges against him; failed to advise him of his right to counsel; failed to adhere to specified standards in judging the issues; denied him meaningful participation in the proceedings including the opportunity to confront and examine witnesses and respond to evidence; and failed to provide a reasoned statement of the decision which they reached.[172]

He also contended that investigators "ignored or misrepresented evidence and leaked confidential information."[173] The court dismissed Popovic's claim, reasoning that he had no due process rights with respect to the agency *investigation*, as opposed to its adjudication.[174] While it is not certain that other courts would agree with this analysis, an individual who is found to have committed research misconduct by ORI, but who successfully challenges ORI's findings at an ALJ hearing, is likely to be unable to make out a viable due process claim.

## 5. OTHER TYPES OF CLAIMS ARISING OUT OF INSTITUTIONAL PROCEEDINGS

### 5.1 State Law Claims

Respondents who feel aggrieved have also brought a variety of nonprocess-related legal claims against institutions and their representatives in connection with research misconduct proceedings. The first group of claims is based in state law, as opposed to federal law. Defamation has proved to be one of the most popular of these claims. However, individuals alleging defamation face several substantial legal obstacles to success. First, the plaintiff claiming defamation must prove that the defendant made a "defamatory" statement, meaning a statement that would expose the plaintiff to "public scorn, hatred, contempt, or ridicule."[175] Reports merely stating the conclusions of institutional research misconduct proceedings are unlikely to meet this standard.[176] Similarly, statements of "pure opinion," as opposed to fact, cannot support a defamation claim.[177] And it is important to understand that a statement is not actionable as defamatory unless it is false.[178]

Even if an individual is able to satisfy these threshold elements of defamation, there are a number of state law "privileges" that may protect statements regarding research misconduct. The "common interest" privilege, for instance, shields "communications made by one person to another upon a subject in which both have an interest."[179] The case of Meena Chandok provides an excellent example of the broad manner in which courts have interpreted this privilege. Chandok brought a defamation suit against Daniel Klessig, the senior scientist in charge of the laboratory where Chandok had formerly worked.[180] After Chandok's resignation, the other scientists in Klessig's lab were unable to replicate certain results that she had reported.[181] Chandok refused to return to the lab to help reproduce the results.[182] Klessig responded to Chandok's stonewalling aggressively by notifying university officials, federal agencies, Chandok's coauthors, and a journal that had published Chandok's results of her possible research misconduct.[183] Klessig also expressed his concerns about Chandok's work to several scientists interested in the relevant field.[184] Ultimately, an institutional committee found the evidence of research misconduct "inconclusive."[185]

The Second Circuit Court of Appeals held that Klessig's statements to university officials and Chandok's coauthors were protected by the common interest privilege.[186] Even Klessig's statements to other scientists were protected because Klessig and those individuals shared a common interest in the relevant field of research.[187]

The court also found a second privilege applicable; namely the privilege for statements made "in the discharge of some public or private duty, legal or moral."[188] Because some of Chandok's research was funded by the federal government, Klessig was under a legal duty to notify the pertinent agencies of his suspicions.[189] Klessig also had a moral obligation to provide this notice, as he was awarded federal funds based on an application that relied heavily on Chandok's data.[190] Similarly, Klessig had a moral obligation to share his concerns with university officials, Chandok's coauthors, and the journals that published Chandok's work.[191]

The privileges for common interest and legal or moral duty sweep broadly, but they are "qualified," that is, not absolute, privileges. Thus, these privileges may be overcome by a showing that the defendant spoke with "malice."[192] In essence, the effect of a qualified privilege is to heighten the level of fault that the plaintiff must prove.

By contrast, an "absolute" privilege precludes liability regardless of the defendant's fault.[193] One such absolute privilege to a claim of defamation is "consent."[194] The case of Roberto Romero illustrates the powerful effect that this privilege can have in the research misconduct context. Romero sued his former colleague at Wayne State University, Irina Buhimschi, for defamation in connection with certain documents submitted by Buhimschi to Yale University, her new employer, and to a medical journal in which she sought publication.[195] These documents accused Romero of various types of unsubstantiated misconduct.[196] Accordingly, while the court hearing the case found them to be "arguably defamatory,"[197] it ruled for Buhimschi, finding that Romero had consented to Buhimschi's publications by initiating related institutional research misconduct proceedings against her at Wayne State.[198]

In addition to defamation, respondents in institutional research misconduct proceedings have brought claims against their accusers for tortious interference with contractual relations. This cause of action, like defamation, contains a number of limitations which have proved problematic for plaintiffs. (While state law governs the legal elements of common law based tort claims like defamation and tortious interference with contractual relations, the basic elements are consistent from one jurisdiction to another.) For example, one court has rejected a tortious interference claim predicated on an institution's notifying the plaintiff's new employer that he was found to have committed research misconduct. The court explained that the plaintiff could not meet the requirement of demonstrating an "unlawful purpose" because his former employer was "privileged to inform future employers of the circumstances surrounding [his] resignation."[199] Similarly, a professor at the university attended by a plaintiff student may have a "privilege to interfere" in the relationship between the plaintiff and the university by reporting alleged misconduct.[200]

One of the other elements of a tortious interference claim is the existence of a business relationship.[201] This element requires "an actual and identifiable understanding or agreement which in all probability would have been completed if the defendant had not interfered."[202] Thus, a student or professor who has applied to, but not yet been accepted by, a new institution will have difficulty formulating a cognizable claim.[203]

While defamation and tortious interference claims have been largely unsuccessful, one type of common law claim could create a trap for the unwary institution. Unprivileged disclosure of institutional reports concerning allegations of research misconduct may give rise to civil liability for invasion of privacy.[204] This is because federal regulations provide that the identities of the parties to research misconduct proceedings must be kept confidential wherever possible.[205] The regulations contemplate disclosure of information regarding ongoing proceedings to a limited universe of people, including respondents, ORI, and, at the election of the institution, complainants.[206] Any additional dissemination could prove problematic. RIOs may wish to utilize written confidentiality agreements to prevent such disclosures.[207]

## 5.2 Federal Claims

In addition to the various state law causes of action discussed above, federal claims may also arise from institutional responses to research misconduct. The First Amendment has been one of the most commonly used vehicles.

Dr Justin Radolf, a tenured professor at the University of Connecticut, brought suit, alleging that his First Amendment right to academic freedom was violated when university officials prevented him from participating in a grant proposal to the Department of Defense (DOD).[208] An institutional committee had previously found that Radolf falsified data in two prior grant proposals, and Radolf later entered into a settlement with ORI, admitting that he had engaged in research misconduct.[209] Under the terms of the settlement agreement, Radolf was placed on probation for 5 years regarding participation in certain activities connected to PHS.[210] In this action, the court ruled against Radolf, declaring that "no court has ever held that a university professor has a First Amendment right of academic freedom to participate in writing any particular grant proposal or performing research under any particular grant."[211] Moreover, the court noted that no evidence was presented that the defendants "denied Dr Radolf participation in the DOD Grant to silence or chill his research on the subject matter of the grant proposal."[212]

Retaliation claims present another First Amendment angle that may be pursued in institutional research misconduct proceedings. One such claim was brought by Ana M. Abreu-Velez against her former employer, the Medical College of Georgia.[213] Abreu-Velez worked as a clinical trials compliance coordinator and research associate.[214] In this capacity, she discovered several "violations of good clinical practices."[215] Abreu-Velez first reported these issues to her supervisors, but, when they failed to take corrective action, she notified an institutional body with the authority to conduct audits of clinical research trials.[216] Abreu-Velez also voiced her concerns to other school officials,[217] and, soon after these events, her employment was terminated.[218]

Abreu-Velez sued, claiming that she was retaliated against by the Medical College for expressing herself regarding the clinical trials.[219] The court began its analysis by noting that "government regulation of a public employee's speech is different from government regulation of the speech of its citizens."[220] "[F]or government employee speech to be protected by the First Amendment, the employee must have (1) spoken as a citizen and (2) addressed matters of public concern."[221] Public employees who "make statements pursuant to their official duties" do not speak as citizens.[222] Applying these principles to the case at hand, the court pointed out that Abreu-Velez "was hired to make sure that data entry for [the] clinical trials was performed in compliance with" the applicable guidelines.[223] Thus, Abreu-Velez's allegations regarding the conduct of the clinical trials were "made as part of her duties."[224] She was speaking as a government employee, not a citizen, and, for this reason, her retaliation claim failed.

The principles articulated by the court in *Abreu-Velez* are widely applicable to public employees who speak out about suspected research misconduct. Under federal regulations, institutions must develop policies and procedures for "[r]eporting and responding to allegations of research misconduct."[225] Thus, when an employee raises concerns about these issues, he or she will likely be acting pursuant to a reporting duty imposed by the institution.[226]

Students at public universities who claim to have been retaliated against for bringing allegations of research misconduct face a less demanding burden. Because students are not government employees, their speech does not fall within the rule relied on by the court in *Abreu-Velez*.[227] "While students' rights are not necessarily coextensive with the rights of adults in other settings, their expressions will generally be protected as long as they do not materially and substantially interfere with the requirements of appropriate discipline or collide with the rights of others."[228]

Federal regulations also require institutions to "[t]ake all reasonable and practical steps to protect the positions and reputations of good faith complainants, witnesses and committee members and protect them from retaliation by respondents and other institutional members."[229] However, once a whistleblower decides to pursue a retaliation claim "through other legal

processes," "ORI will consider that the institutional obligation under the
PHS regulation has been met and ORI will require no further action related
to the whistleblower complaint."[230]

Research misconduct matters can also create issues in federal employment discrimination law under Title VII of the Civil Rights Act of 1964.
A plaintiff bringing suit under this provision has the initial burden of establishing a *prima facie* case of discrimination.[231] Once the plaintiff makes this
showing, the burden shifts to the defendant to provide a "legitimate, non-
discriminatory reason" for the contested employment action.[232] If the
defendant proffers such a legitimate reason, the burden shifts back to the
plaintiff to show that the defendant's reason is merely a pretext for discrimination.[233] An institutional finding of research misconduct constitutes
a legitimate reason for an adverse employment action that an employer may
utilize to shift the burden back to the plaintiff to demonstrate that the stated
reason is pretextual.[234]

Interestingly, plaintiffs in employment discrimination cases may also use
institutional charges of research misconduct to meet their *prima facie* burden.
To establish a *prima facie* case, the plaintiff must show that he or she "suffered
an adverse employment action."[235] At least one court, the Sixth Circuit
Court of Appeals, has held that institutional allegations of research misconduct can, in some circumstances, constitute an adverse employment action
in the related context of a Title VII retaliation claim.[236] In reaching this
result, the court relied on the fact that the research misconduct investigation
at issue "was no run-of-the-mill internal investigation."[237] Before filing
internal charges, the individual bringing the misconduct complaint "emailed
faculty all over the country stating that [the plaintiff] committed research
misconduct."[238] The court's holding was defended on these specific facts,
and it would go too far to say that the simple filing of an institutional
research misconduct complaint, without more, would qualify as an adverse
employment action.[239] But the burden on a plaintiff to make a *prima facie*
showing is fairly light, so assuming some employment consequences can be
tied to the institutional action, the burden will be easily met.

## 6. LITIGATION OVER RECORDS

### 6.1 The Freedom of Information Act

A final area of litigation arising from research misconduct proceedings concerns access to the records generated during those proceedings. The federal
Freedom of Information Act (FOIA) requires government agencies to make

their records available to members of the public upon request.[240] Federal courts have jurisdiction to order agencies to produce improperly withheld records.[241] In order for a private entity to be considered an "agency" subject to FOIA's disclosure requirements, the federal government must have "extensive, detailed, and virtually day-to-day supervision" over that entity.[242] Absent such day-to-day supervision, it is not enough that the federal government owns the physical property on which the entity operates, directs and approves all of the entity's major projects, and provides funding to carry out all of the entity's operations.[243] Moreover, "FOIA applies only to federal and not to state agencies."[244] For these reasons, neither private universities nor their state-funded counterparts will be required by FOIA to disclose records regarding their research misconduct investigations.

HHS, unlike most institutions, constitutes an agency within the meaning of FOIA. However, FOIA allows agencies to withhold documents that fit within certain exemptions. Included on FOIA's list of exempted records are "personnel and medical files and similar files the disclosure of which would constitute a clearly unwarranted invasion of personal privacy,"[245] and "records or information compiled for law enforcement purposes, but only to the extent that the production of such law enforcement records or information...(C) could reasonably be expected to constitute an unwarranted invasion of personal privacy, [or] (D) could reasonably be expected to disclose the identity of a confidential source."[246] ORI takes the position that its records are "primarily protected" by these FOIA exemptions.[247]

At least one federal court has agreed with this proposition. Dr Charles McCutchen submitted a FOIA request to HHS "for a list of all investigations of scientific misconduct" undertaken by ORI, including the names of respondents and complainants.[248] HHS turned over reports relating to investigations which resulted in findings of misconduct, but it withheld information concerning investigations in which no misconduct had been found.[249] McCutchen filed suit in federal court to compel disclosure.[250] Approximately two months later, HHS provided a list of all ORI's closed cases.[251] Omitted from the list, however, were the names of all noninstitutional complainants, the names of all respondents who had not died or been found guilty of misconduct, and the name of any institution whose disclosure might have allowed McCutchen to identify a respondent.[252]

The court began its analysis by explaining that both exemptions 6 and 7(C) "call for a balancing of the privacy interests that would be compromised by disclosure against the public interest in release of the requested information."[253] It found that individuals accused of research misconduct but subsequently exonerated have a significant privacy interest in not having their names disclosed.[254] This is because allegations of research misconduct "carry a stigma and can damage a career."[255] Complainants also have "a strong privacy interest" because, if their identities are disclosed, they might face retaliation.[256] The court rejected the argument that these privacy interests were outweighed by the public interest in ensuring that ORI was conducting comprehensive investigations into suspected research misconduct. It explained, "A mere desire to review how an agency is doing its job, coupled with allegations that it is not, does not create a public interest sufficient to override the privacy interests protected by" the FOIA exemptions.[257] In the case at hand, McCutchen had failed to present any "compelling evidence" that ORI was not doing its job.[258] Ultimately, the District of Columbia Circuit concluded that FOIA exemption 7(C) allowed the agency to withhold the names of respondents and complainants.[259] The court did not address whether the information was also protected by exemption 6 or 7(D).[260]

## 6.2 The Privacy Act

The Privacy Act provides that, with certain exceptions, "[n]o agency shall disclose any record which is contained in a system of records... to any person, or to another agency, except pursuant to a written request by, or with the prior written consent of, the individual to whom the record pertains."[261] To establish a violation of the Privacy Act, a plaintiff must show that the information at issue is a "record" contained within a "system of records."[262] The case of Bernard Fisher, former chairperson of the National Surgical Adjuvant Breast and Bowel Project (NSABP), illustrates how this statute may be implicated in the research misconduct context. In 1990, NSABP, a "consortium of institutions that conducts research on the treatment and prevention of breast and bowel cancer," discovered certain anomalies in data submitted by one of its participating institutions.[263] ORI began investigating Fisher for including the suspect data in published papers.[264] Subsequently, annotations were added to certain computer database files pertaining to articles that incorporated the data.[265] One such annotation read "[scientific misconduct—data to be reanalyzed]."[266] Fisher filed suit in federal court, seeking removal of the annotations.[267]

The court began by noting that, for information to constitute a "record" within the meaning of the Privacy Act, it must "(1) contain the individual's name or other identifying particular and (2) be 'about' the individual."[268] While the database files at issue did contain Fisher's name, they were "about" the articles, not Fisher.[269]

Fisher separately contended that the defendants improperly disclosed information from ORI files about him.[270] The court held that, at the time of the alleged disclosures, ORI's investigatory files did not constitute a "system of records" within the meaning of the Privacy Act.[271] However, ORI now does maintain these files in a system of records.[272] While HHS purports to exempt this system of records from certain provisions of the Privacy Act,[273] the exemption does not extend to 5 U.S.C. Section 552a(b)'s prohibition on the disclosure of records without consent.

Section 552a(b) does not, however, create an absolute bar to nonconsensual disclosure. Rather, it is subject to several exceptions, including one for "routine use."[274] One such routine use identified by HHS provides

> After there is a final HHS/ORI finding of research misconduct, disclosure may be made to professional journals, other publications, news media, other individuals and entities, and the public concerning research misconduct findings and the need to correct or retract research results or reports that have been affected by research misconduct.[275]

Accordingly, individuals found by the agency to have committed research misconduct are listed on the PHS Administrative Action Bulletin Board, available on ORI's website.[276]

## 7. SUMMARY

The above discussion lays out the legal landscape within which scientific research misconduct investigations occur. It can seem at first blush like a complicated tapestry of regulation and case law. But in practice the regulatory requirements imposed by HHS and the ORI are fairly straightforward and provide a step-by-step guide to RIOs charged with handling allegations of misconduct. And because the consequences both to the individual subjects and to the institutions can be so grave (debarment, civil liability, and even criminal liability) there is more than ample incentive for institutions to get it right. And, if institutions do get it right, the above discussion shows that it is quite difficult for subjects to attack the process or the outcome.

As a general matter, several themes emerge from the discussion above, and RIOs conducting investigations would do well to always keep these in mind.

First, RIOs must have a strong working understanding of the regulatory process and their obligations. Strict adherence to that regulatory process is the best insurance against claims regarding mishandling of the alleged misconduct, or claims by subjects regarding abuse of process, breach of contract, and so on. The linchpin of this process is the expeditious and even-handed gathering of all information needed to assess the claim. Facts should be gathered in an open-minded and fair way—and the subject of the investigation should be provided with fair notice of what is being alleged and an opportunity to present his or her side of the story. RIOs should consider using institutional legal counsel during the process to ensure compliance with the regulations and basic due process. The above discussion demonstrates that it is difficult for subjects of a complaint and investigation to successfully challenge the process or outcome of an investigation conducted pursuant to the protocols contained in the HHS/ORI regulatory scheme. But these outcomes are not assured; the successful defense of any challenge to process or outcome depends in large measure on the institution's adherence to protocols outlined in the regulations. Moreover, a well-functioning process for dealing with scientific misconduct allegations—one which strictly adheres to the regulations—will doubtless result in fewer challenges (and therefore lower legal costs) and better outcomes. And at the end of the day, the goal of the legal/regulatory process must be to ensure the integrity of the institution's research by ensuring fairness and accuracy in dealing with allegations of misconduct.

## END NOTES

1. Before ORI was created, its functions were divided between the Office of Scientific Integrity and the Office of Scientific Integrity Review. *See Popovic v. United States*, 997 F. Supp. 672, 679 n.6 (D. Md. 1998), *aff'd*, 175 F.3d 1015 (4th Cir. 1999). In order to avoid confusion, this chapter refers to all three agencies as ORI.
2. *See* National Institutes of Health Revitalization Act of 1993, Pub. L. No. 103-43, Section 161, 107 Stat. 122, 140–141 (codified as amended at 42 U.S.C. Section 289b).
3. HHS amended its regulations governing research misconduct in 2005. *See* Public Health Service Policies on Research Misconduct, 70 Fed. Reg. 28, 370–01 (May 17, 2005). While many of the cases discussed in this chapter involved a prior set of regulations, the cited discussions are equally applicable to the amended regulations unless otherwise noted.
4. 42 C.F.R. Section 93.100(b).
5. Section 93.213.
6. This chapter deals primarily with the regulations promulgated by HHS, which cover institutions that apply for or receive PHS support. Section 93.102. Other agencies, including the National Science Foundation, have promulgated similar regulations concerning research misconduct. See 45 C.F.R. Sections 689.1–689.10. The similarities between the various sets of federal regulations are not the product of mere accident. Indeed, the federal government has adopted a general policy on research misconduct, applicable to research funded by many different federal agencies. *See* Executive Office of the President; Federal Policy on Research Misconduct; Preamble for Research Misconduct Policy, 65 Fed. Reg. 76, 260–01 (December 6, 2000).
7. 42 C.F.R. Section 93.304.
8. Id. A model policy, the Lifespan Hospital System policy on academic misconduct, is included in Appendix 2.
9. 42 C.F.R. Section 93.103.
10. Section 93.103(d).

11. Section 93.104(b).
12. Section 93.104(a).
13. Section 93.319(a).
14. *See Shovlin v. Univ. of Med. & Dentistry of N.J.*, 50 F. Supp. 2d 297, 314 (D.N.J. 1998) ("Even though the federal agency to which the university reported may not have considered duplicate publications to constitute 'misconduct in science', it recognized the University's right to hold such a practice to be unacceptable."); Office of research integrity annual report 2011 [Internet]. Rockville (MD): Department of Health and Human Services (US), Office of Research Integrity; 2011 [cited February 11, 2014]. Available from: http://ori.hhs.gov/images/ddblock/ori_annual_report_2011.pdf [hereinafter Annual Report] (explaining that ORI's definition of plagiarism does not include "efforts by scientists to publish the same data in more than one journal article").
15. *See* Chris B. Pascal, *The Office of Research Integrity*: Experience and Authorities, 35 Hofstra L. Rev. 795 (2006): 805–806.
16. *See* CHPS Consulting, Columbia, MD. Final report. Analysis of institutional policies for responding to allegations of scientific misconduct [Internet]. Rockville (MD): Department of Health and Human Services (US), Office of Research Integrity; 2000 Sep [cited 2014 February 11]. 2.2, 139 p. Contract No.: 282-98-0008. Available from: http://ori.hhs.gov/documents/institutional_policies.pdf.
17. *See* Sample policy and procedures for responding to allegations of research misconduct. [Internet]. Rockville (MD): Department of Health and Human Services (US), Office of Research Integrity; 2005 Jun [cited 2014 February 11]. Available from: http://ori.hhs.gov/sites/default/files/SamplePolicyandProcedures-5-07.pdf [hereinafter Sample Policy].
18. *See Id.*
19. 42 C.F.R. Section 93.307.
20. *See* Sample Policy, supra note 17, at 10.
21. 42 C.F.R. Section 93.307(b).
22. Section 93.317(b).
23. *See* Sample Policy, supra note 17, at 11.
24. 42 C.F.R. Section 93.307(e).
25. *See* Sample Policy, supra note 17, at 12.
26. Section 93.308(a).
27. 42 C.F.R. Section 93.307(f).
28. Section 93.307(d).
29. Section 93.309.
30. Section 93.310.
31. *See* Sample Policy, supra note 17, at 14.
32. 42 C.F.R. Section 93.310.
33. *See* Sample Policy, supra note 17, at 15.
34. 42 C.F.R. Section 93.312(a).
35. *See* Sample Policy, supra note 17, at 28.
36. 42 C.F.R. Section 93.313.
37. Section 93.315.
38. Section 93.316(a).
39. Section 93.316(b).
40. *See* Section 93.413. Similar self-reporting obligations have been used in other high-profile contexts. One high-profile (but non-governmental) example is the National Collegiate Athletic Association ("NCAA"), which requires that member colleges and universities investigate and self-report rules violations. *See* Nat'l Collegiate Athletic Ass'n, Art. 2.8.1, 2012-13 NCAA DIVISION I MANUAL 4 (2012). Failure to

comply with this requirement exposes the member college or university to severe penalties. *See id.* at Art. 19.01.5 (stating that penalties "should be broad and severe if the violation or violations reflect a general disregard for the governing rules").

41. 42 C.F.R. Section 93.400(a).
42. Id.
43. Section 93.403.
44. Id.
45. Section 93.407(a)(11).
46. Section 93.205.
47. Section 93.407(a).
48. Section 93.405(a).
49. Section 93.500.
50. Section 93.406.
51. Id.
52. *See* 5 U.S.C. Section 553.
53. *See* Section 554.
54. *See* Sections 701–706.
55. 42 C.F.R. Section 93.505.
56. Section 93.506(b)(15).
57. Section 93.516.
58. Section 93.219.
59. Section 93.523(a).
60. Section 93.523(b).
61. Id.
62. Section 93.409(a).
63. Annual Report, *supra* note 14, at 16.
64. 31 U.S.C. Section 3729(a)(1)(A). Congress amended Section 3729 in 2009. See Fraud Enforcement and Recovery Act of 2009, Pub. L. No. 111-21, 123 Stat 1617, 1621–1625. The cases discussed all deal with prior versions of the legislation, but the propositions for which they are cited are unaffected by the amendment unless otherwise noted. *See Jones v. Brigham & Women's Hosp.*, 678 F.3d 72, 82 n.17 (1st Cir. 2012) (applying a prior version of the FCA but noting that "[a]pplication of the amended language would not affect the outcome in this case").
65. 31 U.S.C. Section 3729(a)(1)(B).
66. Section 3729(b)(2)(A).
67. Section 3730(b).
68. Section 3729(a)(1).
69. *See, e.g., Jones*, 678 F.3d at 75.
70. *See id.* at 86.
71. 31 U.S.C. Section 3729(a)(1)(B).
72. *See Jones*, 678 F.3d at 86.
73. *U.S. ex rel. Milam v. Regents of Univ. of Cal.*, 912 F. Supp. 868, 873 (D. Md. 1995).
74. Id. at 876.
75. Id. at 875.
76. Id. at 876.
77. Id.
78. Id.
79. Id. at 886; *see also Jones*, 678 F.3d at 87 ("[E]xpressions of opinion, scientific judgments, or statements as to conclusions about which reasonable minds may differ cannot be false."); *U.S. ex rel. Hill v. Univ. of Med. & Dentistry of N.J.*, 448 F. App'x 314, 316 (3d Cir. 2011) (same).
80. *Jones*, 678 F.3d at 87–88.

81. *Luckey v. Baxter Healthcare Corp.*, 2 F. Supp. 2d 1034, 1036–1037 (N.D. Ill. 1998), aff'd, 183 F.3d 730 (7th Cir. 1999).
82. Id. at 1045.
83. Id. at 1046.
84. Id. at 1047.
85. Id.
86. Id. at 1048.
87. *Jones*, 678 F.3d at 91.
88. Id.
89. Id. at 92–93.
90. 42 C.F.R. Section 93.301(b).
91. 31 U.S.C. Section 3729(a)(1).
92. *Luckey*, 2 F. Supp. 2d at 1049.
93. *See* Hill, 448 F. App'x at 316.
94. 31 U.S.C. Section 3729(a)(1)(B); *see also, e.g., Jones*, 678 F.3d at 82.
95. 31 U.S.C. Section 3729(b)(4).
96. *Jones*, 678 F.3d at 93.
97. *See Berge*, U.S. ex rel. Berge v. Bd. of Trs. of the Univ. of Ala., 104 F.3d 1453, 1458 (4th Cir. 1997).
98. *Milam*, 912 F. Supp. at 879–880.
99. Id. at 880.
100. Id.
101. *Brodie*, 796 F. Supp. 2d at 151.
102. 42 C.F.R. Section 93.104.
103. *Milam*, 912 F. Supp. at 880.
104. 31 U.S.C. Section 3730(h)(1).
105. *See Dookeran v. Mercy Hosp. of Pittsburgh*, 281 F.3d 105, 108 (3d Cir. 2002); *Luckey*, 2 F. Supp. 2d at 1050.
106. *Luckey*, 2 F. Supp. 2d at 1050.
107. Id. at 1052.
108. Id.
109. Id. at 1055.
110. *See id.* at 1057.
111. 31 U.S.C. Section 3730(h)(1).
112. *Dookeran*, 281 F.3d at 107.
113. *See id.* at 109.
114. *United States v. Keplinger*, 776 F.2d 678, 683 (7th Cir. 1985).
115. Id. at 684.
116. *See* 18 U.S.C. Section 1341 (prohibiting the use of the mails for the purpose of executing "any scheme or artifice to defraud").
117. *See* Section 1343 (prohibiting the use of interstate wire transmissions for the purpose of executing "any scheme or artifice to defraud").
118. *See* Section 1001 (imposing liability on any person who, within the jurisdiction of the United States government, "knowingly and willfully" "makes any materially false, fictitious, or fraudulent statement or representation").
119. *Keplinger*, 776 F.2d at 685.
120. Id.
121. Id.
122. Id.
123. Id.
124. Id.
125. Id.

126. Id.
127. Id. at 697.
128. Id. at 699.
129. Id. at 689.
130. Id.
131. Id. at 689–690.
132. Id. at 690.
133. Id.
134. *See* Bratislav Stankovi, *Pulp Fiction: Reflections on Scientific Misconduct*, 2004 Wis. L. Rev. 975, 1001 (2004).
135. *Shovlin*, 50 F. Supp. 2d at 300–302.
136. Id. at 302.
137. Id. at 305.
138. Id. at 306.
139. Id. at 306–307.
140. Id. at 308.
141. Id.
142. *See* 32 Charles Alan Wright and Charles H. Koch, Jr., Federal Practice and Procedure Judicial Review Section 8128 (1st ed.).
143. Note, however, that there are other obstacles to suits against public universities and their employees, namely sovereign and qualified immunity, which are discussed below at p. 27.
144. *See Radolf v. Univ of Conn*, 364 F. Supp. 2d 204, 210 (D. Conn. 2005); *Needleman v. Healy*, CIV.A. 92-749, 1996 WL 33491149, at *4 (W.D. Pa. May 22, 1996), *aff'd sub nom. Needleman v. Varmus*, 127 F.3d 1096 (3d Cir. 1997).
145. *Shovlin*, 50 F. Supp. 2d at 316; *see also, e.g., Radolf*, F. Supp. 2d at 211-13, 218-21, 226–228 (finding no property interest in reinstatement of a voluntarily relinquished title, participation in a grant proposal that the plaintiff had not previously worked on, or certain rights of a tenured professor).
146. *Shovlin*, 50 F. Supp. 2d at 316; *see also Popovic*, 997 F. Supp. at 680 ("[T]here is no constitutional protection of a person's interest in his or her reputation alone."); *Needleman*, 1996 WL 33491149, at *5 ("It is well established that reputation alone is not a constitutionally protected interest, but that loss of reputation must be accompanied by a change or extinguishment of a right or status guaranteed by state law or the United States Constitution.").
147. *See Radolf*, 364 F. Supp. 2d at 210.
148. *See Qvyjt v. Lin, 932 F. Supp. 1100, 1106 (N.D. Ill. 1996)*.
149. *See Radolf*, 364 F. Supp. 2d at 213.
150. *See id.*
151. Id. (emphasis added) (internal quotation marks omitted) (alterations in original).
152. *See Needleman*, 1996 WL 33491149, at *4.
153. *Kerr v. Johns Hopkins Univ*, C.I.V. L-10-3294, 2011 WL 4072437, at *1 (D. Md. September 12, 2011), aff'd, 473 F. App'x 246 (4th Cir. 2012).
154. Id. at *4–5.
155. Id. at *5 (alteration in original).
156. *See Sirpal v. Univ of Miami*, 09-22662-CIV, 2011 WL 3101791, at *15 (S.D. Fla. July 25, 2011), aff'd, 11-15210, 2013 WL 599730 (11th Cir. February 19, 2013).
157. *Sadwick v. Univ of Utah*, 2:00-CV-412C, 2001 WL 741285, at *1 (D. Utah April 16, 2001).
158. Id. at *1–2.
159. Id. at *2.
160. Id. at *6.

161. Id.
162. Id at *4 (emphasis added); see also Radolf, 364 F. Supp. 2d at 218.
163. Suits brought against state universities under the False Claims Act ("FCA"), discussed above at p. 19–20, are not barred by the Eleventh Amendment. This is because the federal government, which is the real party in interest in an FCA suit, "may sue states in federal court." U.S. ex rel. Berge v. Bd. of Trs. of the Univ. of Ala., 104 F.3d 1453, 1458 (4th Cir. 1997).
164. See, e.g., Radolf, 364 F. Supp. 2d at 209.
165. Brodie v. U.S. Dep't of Health & Human Servs., 796 F. Supp. 2d 145, 148 (D.D.C. 2011).
166. Id. at 150.
167. Id. at 156.
168. Id. at 157.
169. See Abbs v. Sullivan, 963 F.2d 918, 925–927 (7th Cir. 1992).
170. Popovic, 997 F. Supp at 675.
171. Id.
172. Id.
173. Id.
174. Id at 678–679.
175. Kerr, 2011 WL 4072437, at *6 (internal quotation marks omitted); see also Sirpal, 2011 WL 3101791, at *16 ("[W]ords are defamatory when they charge a person with an infamous crime or tend to subject one to hatred, disgust, ridicule, contempt or disgrace or tend to injure one in one's business or reputation." (internal quotation marks omitted)).
176. See Kerr, 2011 WL 4072437, at *6.
177. See Sirpal, 2011 WL 3101791, at *16.
178. See id. at *17.
179. Chao v. Mount Sinai Hosp, 11-1328-CV, 2012 WL 1292757, at *1 (2d Cir. April 17, 2012).
180. Chandok v. Klessig, 632 F.3d 803, 805–806 (2d Cir. 2011).
181. Id. at 806.
182. Id. at 807.
183. Id at 808–809.
184. Id. at 809.
185. Id.
186. Id. at 817.
187. Id. Other courts have interpreted this privilege in a similarly expansive manner. One court has broadly declared that statements made "as part of an investigation" into allegations of research misconduct are protected by the common interest privilege. See Sirpal, 2011 WL 3101791, at *18. This understanding of the privilege will pose a significant obstacle to defamation claims arising from institutional research misconduct proceedings.
188. Chandok, 632 F.3d at 814.
189. Id. at 816.
190. Id.
191. Id.
192. See id at 815; Sirpal, 2011 WL 3101791, at *16.
193. See Romero v. Buhimschi, 2:06-CV-10859, 2009 WL 2477556, at *8 n.7 (E.D. Mich. Aug. 10, 2009), aff'd, 396 F. App'x 224 (6th Cir. 2010).
194. See Romero, 396 F. App'x at 236.
195. Romero, 2009 WL 2477556, at *1.
196. Id.
197. Id. at *5.

198. Id. at *7.
199. *Kerr*, 2011 WL 4072437, at *7.
200. See *Sirpal*, 2011 WL 3101791, at *20.
201. See *id*. at *19.
202. Id. at *20 (internal quotation marks omitted).
203. See *id*. at *20–21.
204. See *Arroyo v. Rosen*, 648 A.2d 1074, 1080 (Md. Ct. Spec. App. 1994).
205. 42 C.F.R. Section 93.108.
206. Sections 93.308(b), 93.312(b).
207. See Sample Policy, supra note 17, at 7.
208. *Radolf*, 364 F. Supp. 2d at 215.
209. Id. at 208.
210. Id.
211. Id. at 216.
212. Id. at 217.
213. *Abreu-Velez v. Bd of Regents of Univ Sys of Ga*, CV 105-186, 2009 WL 362926, at *1 (S.D. Ga. Feb. 12, 2009), *aff'd*, 328 F. App'x 611 (11th Cir. 2009).
214. Id. at *2.
215. Id.
216. Id. at *3.
217. Id.
218. Id. at *4.
219. Id. at *7.
220. Id.
221. Id.
222. Id (quoting *Garcetti v. Ceballos*, 547 U.S. 410, 421 (2006)).
223. Id. at *8.
224. Id.
225. 42 C.F.R. Section 93.101(d).
226. See Sample Policy, *supra* note 17, at 6 ("All institutional members will report observed, suspected, or apparent research misconduct to the RIO").
227. See *Qvyjt*, 932 F. Supp at 1108.
228. Id. at 1108-09 (internal quotation marks omitted).
229. 42 C.F.R. Section 93.300(d).
230. Office of research integrity annual report 2010 [Internet]. Rockville (MD): Department of Health and Human Services (US), Office of Research Integrity; 2010 [cited February 11, 2014].
231. See *Chao*, 2012 WL 1292757, at *2.
232. See *id*. (internal quotation marks omitted).
233. See *id*.
234. See *id*. at *3; *Sirpal*, 2011 WL 3101791, at *13.
235. See *Chao*, 2012 WL 1292757, at *2.
236. *Szeinbach v. Ohio State Univ.*, 11-3002, 2012 WL 3264398, at *5 (6th Cir. Aug. 10, 2012).
237. Id. at *4.
238. Id. at *5.
239. *Cf Radolf*, 364 F. Supp. 2d at 224–226 (holding that merely being the subject of an internal research misconduct investigation did not constitute an adverse employment action capable of supporting a First Amendment retaliation claim).
240. 5 U.S.C. Section 552(a)(3).
241. Section 552(a)(4)(B).

242. *Reich v. U.S. Dep't of Energy*, 784 F. Supp. 2d 15, 20 (D. Mass. 2011) (internal quotation marks omitted).
243. *See id.*
244. *Grand Cent P'ship, Inc v. Cuomo*, 166 F.3d 473, 484 (2d Cir. 1999).
245. 5 U.S.C. Section 552(b)(6).
246. Section 552(b)(7)(C)-(D).
247. *See* Annual Report, supra note 14, at 45.
248. *McCutchen v. U.S. Dep't of Health & Human Servs.*, 30 F.3d 183, 184 (D.C. Cir. 1994).
249. Id at 185–186.
250. Id. at 186.
251. Id.
252. Id.
253. Id. at 185 (internal quotation marks omitted).
254. Id. at 188.
255. Id. at 187.
256. Id. at 189.
257. Id. at 188.
258. Id. at 189.
259. Id. at 189–190.
260. Id. at 190.
261. 5 U.S.C. Section 552a(b).
262. *See Fisher v. Nat'l Insts of Health*, 934 F. Supp. 464, 468 (D.D.C. 1996), *aff'd*, 107 F.3d 922 (D.C. Cir. 1996).
263. Id.
264. Id. at 467.
265. Id.
266. Id.
267. Id.
268. Id. at 468.
269. Id. at 469.
270. Id. at 468.
271. Id. at 473.
272. *See* Privacy Act; Exempt System, 59 Fed. Reg. 36, 717–01 (July 19, 1994).
273. *See* Office of Research Integrity; Privacy Act of 1974; Report of an Altered System of Records, 74 Fed. Reg. 44, 847–01 (August 31, 2009).
274. 5 U.S.C. Section 552a(b)(3).
275. Office of Research Integrity; Privacy Act of 1974; Report of an Altered System of Records, 74 Fed. Reg. at 44,847–901.
276. *See* U.S. Department of Health and Human Service, Office of Research Integrity. PHS administrative action report [Internet]. Rockville (MD): U.S. Department of Health and Human Service; [updated April 2, 2014; cited 2014 April 2]; [about 2 screens]. Available from: http://ori.hhs.gov/ORI_PHS_alert.html.

# Case studies

## A Note on the Presentation of Nine Research Integrity Case Studies

The following nine case studies have been prepared by the authors, drawing on their experiences with specific cases managed by them and others at various academic medical institutions over a number of years.

In order to protect the confidentiality of all individuals referenced within these case studies, all identifying information have been intentionally altered. Specifically: (1) these cases are presented in a standardized, structured format, with consistently applied style, to obscure assignment of authorship to any one of the authors of this volume; (2) all nine chapters are presented in a pseudo-randomized order; (3) all identifying features of individuals (e.g., names, sex, nationality, race) have been altered; (4) specific topical areas for each case (i.e., field of scientific research or publication) have been changed; (5) important dates have been modified; and (6) for cases that include portions of transcribed interviews, such interviews have been fictionalized and based on heavily redacted source material.

CHAPTER 3

# Question of Data Manipulation in Failure to Replicate Findings in Published Manuscript

## 1. PRESENTING COMPLAINT

In the spring of 2011, Ms Morgan, a graduate student in neurobiology, sought consultation about a recently published manuscript reporting on the impact of genetic ablation of a specific protein X on synaptic morphology and dendritic spine density. The manuscript was published in a high-impact journal and had been through three revisions prior to final acceptance for publication. Early in her work in Dr Zimmerman's lab, Ms Morgan had been unable to replicate the findings reported in Dr Foster's manuscript. In her effort to accomplish this replication, Ms Morgan became aware of significant differences between the published methods and those suggested in verbal conversations with Dr Foster. After consultation with other faculty advisors, Ms Morgan sought the advice of the research integrity officer (RIO).

## 2. BACKGROUND AND HISTORY

Ms Morgan was a talented, productive doctoral student and worked well with Dr Zimmerman. Indeed, she had been included as an author on four manuscripts from the lab in the 2 years since joining the research group. She was scrupulous in her procedural documentation and in her attention to detail. Although a junior investigator with only a few years of experience in neurobiology, her colleagues respected her ability to accomplish difficult experiments and to effect procedures that even more experienced lab technicians could not replicate. Thus, her failure to replicate Dr Foster's work was notable. Although Ms Morgan had moved on to the specific area of work that was the focus of her dissertation, she remained somewhat preoccupied with thinking about the discrepancies between the procedures

*The Management of Scientific Integrity within Academic Medical Centers*
http://dx.doi.org/10.1016/B978-0-12-405198-0.00003-5

reported in the 2011 published paper and the methods Dr Foster had described to her. Further, when she had brought her concerns to Dr Zimmerman over a year ago, he had told her not to worry about it and to move on with her own dissertation research. Although their working relationship was consistently positive, Ms Morgan experienced Dr Zimmerman as irritated and abrupt in response to her concerns about this particular experiment. The senior postdoctoral fellow in the laboratory had also advised Ms Morgan just to focus on her own work and not pursue her concerns. Careful as she was, Ms Morgan could not contain her suspicion that this advice from her mentor and senior postdoctoral colleague was covering up some procedural irregularities that they did not want to review. Still she continued with her own research until a required lecture on research ethics and scientific misconduct reminded her of a shared responsibility of members of the scientific community to ensure the integrity of the scientific process and of the published scientific record. After consulting with a trusted senior advisor who also knew and respected Dr Zimmerman, Ms Morgan decided to consult with the RIO of the graduate school.

Dr Zimmerman was a respected, highly productive, and innovative neuroscientist. From early in his career, he had published consistently in high-impact journals including *Nature*, *PNAS*, and *Science*. His laboratory had made significant contributions to understanding the genetic regulation of synaptic morphology and function. His laboratory's reputation ensured a steady stream of talented pre- and postdoctoral students, and many junior faculty around the world got their start in Dr Zimmerman's laboratory. Dr Zimmerman valued his role as a teacher and mentor and, as much as possible, tried to provide individual attention to each of his mentees. But as his lab had grown, he had come to rely more and more on senior postdoctoral fellows and junior faculty to oversee students and more junior colleagues in the lab. Increasingly, he was distant from the day-to-day work in the lab, and it was now far less common than in the beginning of his career that Dr Zimmerman was able to review raw data and procedures with his mentees. A year ago, Dr Zimmerman had an opportunity to design and implement an interdisciplinary program in genetic regulation of synaptic structure and function. His lab joined with four other labs, two local and two international, in a large, translational neuroscience program. While bringing more resources to his lab, this opportunity also took Dr Zimmerman even further from the day-to-day work in his lab. He trusted his students and postdoctoral fellows and believed each followed his own standards of scientific integrity and responsibility.

Dr Foster was one of the more talented postdoctoral fellows to join Dr Zimmerman's laboratory. Trained at a lab abroad by a well-known leader in the field who was senior to Dr Zimmerman, Dr Foster was excited to join such a vibrant laboratory for he was eager to develop a sufficiently strong portfolio so he could move quickly to a faculty job and establish his own laboratory. Dr Foster brought ideas for work from his previous labs that were consonant with some methods used by Dr Zimmerman's team. His questions regarding the neuroregulatory functions of protein X were synergistic with a number of experiments ongoing in Dr Zimmerman's lab. He accomplished his work with the support of a 2-year NIH grant awarded to Dr Zimmerman as principal investigator.

## 3. RELEVANT EVIDENCE

When Ms Morgan approached Dr Foster about his methods, she was surprised to learn of what she considered some critical shortcuts in research methods. The experiment required detailed counting of dendritic spines. Dr Foster revealed that he had selected two classes of dendritic spines to count and had not counted two other classes based on previous work from the lab about the specificity of the effect of protein X. Although perhaps an appropriate scientific decision, the reasons for this selective counting were not fully explicated in the published manuscript. The selection of classes of dendritic spines might well provide very different results regarding the effects of the genetic ablation of protein X. Further, when methods require counting neuronal structures, investigators might photograph images in the microscopic field so as to allow reliability checks on their counts. When discussing Ms Morgan's failure to replicate her findings, Dr Foster was not able to provide images of his work for he had not photographed the fields as he counted the structures.

When Ms Morgan consulted with the RIO, she expressed both her concerns about Dr Foster's variances from what she perceived as a standard procedure, that is, to count all classes of a particular neuronal structure, and the unwillingness of her mentor and the postdoctoral lab manager to review her concerns. The RIO consulted with two senior faculty, both experts in the specific scientific fields of the published manuscripts, and asked these two faculty a hypothetical question: If Ms Morgan's allegations were true, might they constitute scientific misconduct? Based on the advice from these two faculty members, the RIO proceeded to open an inquiry with three senior faculty members from the graduate school. On notification of the inquiry procedures, Dr Foster quickly responded by sending drafts of the manuscript and supplementary materials. Dr Zimmerman had no additional materials to supply but was also quickly responsive, concerned, and strongly supportive of Dr Foster's scientific integrity and research skills. Understanding that Ms Morgan had brought the allegations to the attention of the RIO, Dr Zimmerman was initially surprised but then concerned that Ms Morgan had not persisted in her efforts to speak with him.

## 4. CASE MANAGEMENT AND ADJUDICATION

The faculty committee reviewing Ms Morgan's allegations spoke at length with Ms Morgan, Dr Zimmerman, and Dr Foster. Dr Foster took the inquiry process seriously and provided a detailed description of his experimental procedures. He also supplied all the supplemental information he had provided the journal in the submission and revision process. In terms of the allegation that Dr Foster had selectively narrowed classes of dendritic spines, he counted in favor of those supporting his hypothesis. Dr Foster explained that he had indeed counted only the spine classes that previous work had suggested were most regulated by protein X. Although he had not clearly cited his reasons for the selections of what to count, he had commented in the discussion of the paper that protein X appeared highly specific in its regulatory effects even within subtypes of specific neural structures. Further, in a response to reviewers who asked about the other classes of dendritic spines not apparently counted, Dr Foster and Dr Zimmerman had stated their opinion that the other classes had been shown in previous work not related to protein X. The editor and reviewers appeared satisfied with their response and did not require further explication. The inquiry committee concluded that Dr Foster's decisions and actions regarding which classes of spines to count in the visual field did not constitute intentional falsification, deception, or gross negligence but were rather potential points of scientific discussion and debate.

Regarding the allegation of failure to make images of the relevant fields in which the dendritic spines were counted, Dr Foster noted that there were images reproduced in the published paper. He also provided additional images from his files when requested by the committee. However, he noted that these images required considerable digital storage space and hence, his usual practice was to take samples of images but not to capture every image he counted. When he had moved to his faculty job and after the manuscript was accepted for publication, he had also deleted stored images to free up space on the laboratory computer system. He also described his usual practice of counting without verification by imaging the selection or having another individual count with her. Moreover, in order to count neural structures, Dr Foster needed to focus up and down and a single image of the visual field he was counting would not have fully captured all the spines, especially the smaller class of spines relevant to the study, in that specific visual field. Dr Foster provided the committee copies of his data worksheets on which he recorded his counts. The committee also consulted with other scientists familiar with the experiments and methods and was told Dr Foster's approach was accepted in the field. Hence, the committee concluded that Dr Foster's actions in regard to the second concern also did not constitute gross negligence or data falsification or fabrication.

## 5. SYNTHESIS AND FORMULATION

Ms Morgan's concerns reflected her own careful approach to methods, her wish to be responsible in her new scientific career, and her interpretation of what is appropriate scientific method. Her own failure to replicate Dr Foster's findings raised her concern and then her mentor's unwillingness to engage in a more detailed discussion heightened her worry. The senior postdoctoral fellow's caution only further heightened her worry that somehow by not reporting her concerns she was complicit in wrongdoing. However, Dr Foster was detailed in his explanations and the committee understood that he was following standard practice in his field. Hence, while the finding was not replicable in Ms Morgan's hands, there was no suggestion that Dr Foster had intended to deceive or been grossly negligent.

## 6. RESOLUTION

The RIO acting on behalf of the dean of the graduate school informed both Dr Foster and Dr Zimmerman that the inquiry committee did not recommend any further investigation into the allegations brought forward by Ms Morgan. Ms Morgan was similarly informed. Dr Zimmerman continued his mentorship of Ms Morgan in her dissertation research, and both were able to find time to discuss in more detail questions of replicability and variations in methods and procedures across fields. Further, while Dr Zimmerman was initially surprised, even angry, that Ms Morgan took her concerns to the RIO, he was also able to understand her decision from her point of view and to institute a regular time to meet with her and also to make clear to his lab that he was open to discussion about any concerns emanating from his team.

## 7. COMMENTARY

Failure to replicate is not proof of data manipulation, fabrication, or even gross negligence. The scientific literature is replete with first publications that are never replicated in subsequent experiments even in the same laboratory and often in the hands of the same investigator (Ioannidis, 2005). Indeed, the biotech company Amgen assembled a team to attempt to reproduce the findings of 53 "landmark" articles in cancer research published by reputable labs in top journals, but only 6 of 53 (11%) studies were replicable (Begley and Ellis, 2012). At the same time, the pressures for scientific productivity and continued "discovery" may not reward laboratories' efforts to replicate their own findings and thus, the literature is biased toward nonreplicable findings (Fanelli, 2010, 2012). Still, from the scientific integrity point of view, simple failure to replicate is not synonymous with scientific misconduct.

Accepted standards for experimental procedures also vary across fields and even within fields. What is proper or accepted in one setting for ensuring accuracy and reliability may not be the same in another, and these standards vary as technology evolves. Thus, there is not one standard for proper scientific procedures that can be universally applied when questions of improper scientific conduct arise. The standards of the field in question become central to the inquiry and may change with time. That in turn may contribute to different adjudications of similar allegations from different eras.

Finally, this case also illustrates the importance of open communication with junior scientists who are learning their craft and who may be especially concerned about variations in procedures and/or their own failure to replicate findings by more senior colleagues. Simple instruction in the standards of scientific conduct is important but not sufficient. Ms Morgan understood the principles of scientific conduct but did not have the opportunity to have a nuanced and thoughtful discussion about her concerns with her more senior colleagues. With more opportunity to express her concerns and more openness to hearing her worries, there may have also evolved opportunities to improve lab procedures and learning among all members of the team.

## QUESTIONS FOR DISCUSSION

1. Was Ms Morgan's decision to go the RIO understandable? What other avenues might she have pursued?
2. Was the RIO's decision to begin an inquiry process reasonable? Were there other avenues to pursue? How did the fact that the work was accomplished with National Institutes of Health (NIH) funding influence the decision-making process?
3. Communication issues are very common across and within research labs. How might Dr Zimmerman have ensured open communication even as he became less available on a day-to-day basis in his lab?
4. Ms Morgan's decision to contact her RIO and allege misconduct stems from a misunderstanding between who needs to understand the sources of potential discrepancies and how clearly/fully that knowledge needs to be relayed to others. What can be done to encourage the necessary communication to ensure those with concerns about potential research misconduct and those who better understand the research can resolve issues more efficiently and without initiating an unnecessary investigation?
5. In this case, a required ethics lecture prompted Ms Morgan to report the suspected misconduct. It seems that, absent this lecture, the issue may never have been reported. Should all institutions hold lectures like this on a regular basis?
6. Should institutions also provide training to lab supervisors like Dr Zimmerman (and his senior postdoctoral fellow)? It seems likely that, had Dr Zimmerman been sufficiently responsive to Ms Morgan's concerns, the situation could have been resolved without resorting to the formal inquiry process.
7. How could Dr Zimmerman have more effectively supervised his mentees? Would it have been realistic for him to review all of their raw data?
8. Before initiating an inquiry, the RIO consulted two experts in the relevant field. Is this a step that should be taken in every case to ensure that a mere scientific misunderstanding is not elevated into a research misconduct investigation?
9. Is it possible that a more thorough citation by Dr Foster regarding the images used in the paper could have avoided the problem?
10. Do you think the term "allegation" is too aggressive for some types of concerns? Could a more neutral term avoid some of the hard feelings and resentment that can result from these cases?

# REFERENCES

Begley, C.G., Ellis, L.M., 2012. Drug development: raise standards for preclinical cancer research. Nature 483, 531–533.

Fanelli, D., 2010. "Positive" results increase down the hierarchy of the sciences. PLoS ONE 5 (4), e10068.

Fanelli, D., 2012. Negative results are disappearing from most disciplines and countries. Scientometrics 90, 891–904.

Ioannidis, J.P.A., 2005. Why most published research findings are false. PLoS Med 2 (8), e124.

CHAPTER 4

# Violation of Institutional and Public Health Service Policies Governing the Care and Use of Animals in Research: Cultural Barriers in the Academic Workplace

## 1. PRESENTING COMPLAINT

On August 4, 2010, Dr Lorinda Taylor submitted an incident report to the Institutional Animal Care and Use Committee (IACUC) at the medical center. According to her report, Dr Taylor had been notified of inappropriate use of the mice under her responsibility; mice that had originally belonged to Dr Thomas Dekker, an orthopedics researcher who had left the hospital for employment at a medical center in a neighboring state. Dr Taylor assumed responsibility for Dekker's protocol in January of 2010, with the understanding that his mouse colony would only be maintained until Dr Dekker could breed a new colony (of the same genetic strain) at his new institution. Since no experimentation, breeding, or other manipulation would be required, Dr Taylor separated all breeding pairs and removed Dr Dekker's name from the cages, in compliance with the IACUC protocol and the hospital's animal care facility policies.

The unusual activity within the colony was first detected at the end of June 2010, at which time Dr Taylor was alerted to the fact that a mouse had been injected on June 26 and found dead 4 days later. Dr Taylor reported the incident to the chairperson of the Animal Welfare Committee, and she kept watch on the colony for more evidence of tampering. Upon discovering further evidence of unauthorized rearrangement of animals between cages, tail snips for genotyping, and more injections, Dr Taylor submitted her incident report to the IACUC.

*The Management of Scientific Integrity within Academic Medical Centers*
http://dx.doi.org/10.1016/B978-0-12-405198-0.00004-7

## 2. BACKGROUND AND HISTORY

In 2006 Dr Thomas G. Dekker established a breeding colony of mice under his IACUC-approved project, on the topic of cartilage degeneration in osteoporosis. Dr Akil Gamal, his postdoctoral fellow, and Ms Eshe Fathi, a lab technician, both worked on Dekker's study while employed at the hospital. Dr Dekker and Dr Gamal left the institution, together, for employment at a hospital in a neighboring state, in July of 2009. At that time, Dr Dekker transferred a number of breeding pairs to his new research facility and requested that the existing colony be maintained locally until his transgenic mouse line was firmly established at the new facility. In August of 2009, Dr Taylor assumed responsibility for the colony to continue some of the experiments and to maintain the breeding colony at her colleague's request. Ms Fathi remained at the hospital, but she was transferred to a different laboratory under another principal investigator. Dr Taylor had collaborated with Dr Dekker on previous projects and agreed to assume responsibility for the colony with the understanding that no experiments and no procedures would be done with the mice, and that they would be sacrificed as soon as his new colony was self-sufficient. In January and February of 2010, Dr Taylor sacrificed all nongenetically modified mice that were no longer necessary for Dr Dekker's experiments to save extra maintenance charges, since Dr Dekker was responsible for the cost of maintaining the colony. Because Dr Gamal requested that he be able to finish some work with the animals still being housed locally, Dr Taylor granted Dr Gamal extended privileges to the lab for (only) the month of February 2010, with the caution that "he was not allowed to initiate any new studies alone and that [Dr Taylor] would be able to help him with any issues".

On February 6, 2010, the IACUC coordinator notified Dr Taylor that a progress report and request for renewal were necessary if the protocol was to continue beyond March 1, 2010. After several attempts to get in touch with Dr Dekker for instructions, Dr Taylor submitted the necessary documents filled out to the best of her understanding, and the protocol was renewed. On March 7, 2010, Dr Dekker requested that the protocol for the colony be renewed for 500 mice, with the assumption that establishing a successful colony at his new institution would take no longer than 2–3 months. Dr Taylor agreed to this request, and the colony was maintained for this purpose.

On June 30, 2010, Dr Taylor received notice from the animal care facility that one of the colony's mice had been found dead. The card on the cage listed the birth date of the mouse as March 11, its arrival at the facility on June 26, and a record of an injection administered to the mouse on the same day. Dr Taylor had ordered 45 transgenic male mice, born on March 11 and received from the lab's supplier, on June 26 for a separate experiment of her own. The same dates on the order matched the dates on the cage card for the mouse found dead on June 30, but no mice from that shipment were intended to be used in the protocol for Dr Dekker's previously established colony.

With further observation and genetic testing, Dr Taylor confirmed that mice intended for her own use, and within her own separate colony, were being removed and used for alternate purposes. One of the mice in a cage labeled "Akil" was determined to be a Taylor mouse based on blood testing. In another instance, a cage card noted injections, unauthorized by Dr Taylor, to the mice inside. In a number of other cages, there was evidence of genotyping from tail snips. Finally, in one cage, there was evidence of breeding within the facility, which was highly contradictory to the "maintenance-only" responsibility that Dr Taylor had agreed to. Dr Taylor submitted the incident report to the IACUC committee as soon as she had collected evidence of the genetic testing, breeding, and experimentation that had occurred without her knowledge.

## 3. RELEVANT EVIDENCE

On July 14, 2010, Dr Taylor examined a number of her mice that were to be used in an experiment the following day. She discovered five mice missing from one of her cages and a new cage of five mice near the cage in question. A week later, Dr Taylor looked in again on the mice to be used in an experiment the following day. She found that the new cage that had contained five mice the previous week now only contained one, and was labeled with a corresponding cage card with the name "Akil". The handwriting on the "Akil" cage card was verified by several individuals familiar with Akil Gamal's handwriting to be Dr Gamal's. The cage which had previously lost five mice was once again at its regular 10 mouse capacity.

The following day, Dr Taylor performed a genetic test on the mouse in the "Akil" cage and confirmed that it was one of the mice ordered for another experiment in the Taylor lab. A total of seven other cages showed evidence of manipulation after January 1, 2010, including weaning and tail snips for genotyping.

In addition to the evidence left in the animal care facility, numerous employees confirmed that they had seen Dr Gamal return to the animal care facility after his employment had ended, and after he should have returned both his hospital I.D. badge and keys. As the final enquiry report eventually noted, "It does not appear that Dr Gamal tried to hide his comings and goings, but he was not forthcoming in admitting that he was returning to conduct research. Rather, he stated that he came to attend a class, which might have occurred coincidentally with the incident under investigation". Thus, the evidence does point to Dr Gamal's unauthorized presence and actions with the Taylor lab animals well after both his research with Dr Dekker, and after the 1 month extension that he received from Dr Taylor, had ended.

## 4. CASE MANAGEMENT AND ADJUDICATION

Within a week after Dr Taylor filed her incident report, the institution's Research Integrity Officer (RIO) asked the Chairperson of the IACUC, Dr Jim Stevens, to both form and to chair a subcommittee that was charged with conducting "an enquiry of the circumstances and facts related to [the] complaint filed by Dr Lorinda Taylor on August 4, 2010". The subcommittee consisted of faculty from several hospitals and laboratories within the medical system, as well as faculty from the academic campus of the university. The subcommittee took immediate measures to preserve all hospital property while the investigation was in progress and began by moving Dr Taylor's mice to a different animal housing room that would be accessible only by Dr Taylor and animal care staff. In addition, the remainder of the mouse colony in question was euthanized at the request of Drs Taylor and Dekker.

Based on the multiple failures and misunderstandings in communication between the principal players in the events in question, the subcommittee's first task was to sort out the actions of each individual. They acquired a basic timeline through interviews with some of the Lifespan employees involved in the events, including Dr Taylor, Ms Eshe Fathi, the manager of the animal facility, as well as a review of all relevant correspondence between the central protagonists in this case.

The interviews with Ms Fathi were particularly revealing for this investigation, since her role in the events was unclear from the contradicting e-mails that the subcommittee had received. Ms Fathi was interviewed by the Chair of the investigative committee, and one member who is fluent in Ms Fathi's native language (Arabic), in an attempt to minimize any unnecessary feelings of pressure or anxiety that Ms Fathi might feel. Selected passages from the committee's interview with Ms Fathi, on October 15, 2010, follow (JS = Jim Stevens, EF = Eshe Fathi):

JS: The incident occurred sometime in June of this year, and you stopped working with Dr Dekker last year. Is this correct?

EF: Yes.

JS: When did you stop working for Dr Dekker?

EF: End of July, 2009.

JS: When you worked for Dr Dekker, what were your responsibilities?

EF: I did genotyping and breeding in the mouse colony.

JS: When Dr Dekker left the institution, what lab employed you next?

EF: I started to work for a urology basic science lab in the same building.

JS: Do you know who took over the responsibility of looking after the mice in the colony formerly managed by the Dekker lab?

EF: I don't know.

JS: It appears that those mice continued to be bred and they continued to be genotyped. Both Drs Dekker and Gamal had left the institution, and this is why we are now confused. Who was doing that work?

EF: I don't know.

JS: Was there any time when Dr Lorinda Taylor told you that the project was finished and not to do any more work with those animals?

EF: No.

JS: Did you ever have any contact with Dr Dekker after he left? Did he ever call you?

EF: Akil Gamal called me and asked me to help him to check the mice, and I did not have time because I was busy. (EF then reported, through the Arabic-speaking committee member, that Dr Gamal had asked her to cull pups, and to call him when there were pups, but she never did because she was too busy with her daily work.)

JS: Do you recall when that was?

EF: Three months after Dr Dekker left, maybe September or October.

JS: We have some cards from the animals. Somebody was doing some work with these mice and I would like to show you these cards. Is this your handwriting?

EF: No.

JS: When did you find out that there was a problem here? Did somebody call you? Did Dr Taylor or someone contact you and ask if you knew anything about this incident?

EF: Nobody called me.

JS: Until the committee contacted you, Dr Taylor had not contacted you?

EF: Dr Taylor sent an email to me and she wanted me to call her back, but I never called her because I don't have time.

JS: How often did you go into the animal room?

EF: Every day I checked the mice for the urology lab I work in. (EF then reported, through the Arabic-speaking committee member, that she never saw anyone in the colony room besides the animal care facility staff.)

JS: Dr Taylor said that somebody was continuing to breed her mice after you stopped, because new pups were being separated and tail snips were being done, but it seems that you were not doing these. Do you have any idea who might have been continuing to use those animals for research?

EF: I have no idea. I don't know who would do that.

JS then showed another cage card to Ms Fathi and pointed out the handwriting that matched other cards, and other handwriting that did not match. JS commented that it looked to be from someone else; that is, it looked as though two different people had written on the cage cards found with the colony in question.

JS: When Drs Dekker and Gamal were here, you were the only lab technician conducting the breeding? Or was there someone else?

EF: Only myself.

JS: Was there anyone else who worked in that lab besides yourself?

EF: (EF then reported, through the Arabic-speaking committee member, that a graduate student, named Stephanie, also had access to the colony....EF did not recall her last name. She worked under Dr Dekker, and she left the lab when Dr Dekker resigned.)

JS: We are confused because somebody did this and we don't know if Dr Dekker or Dr Gamal traveled back to this institution to run new experiments.

EF: Dr Dekker's group had 4 people. All three others left, other than myself, when Dr Dekker resigned (one of these employees was Chinese, and he returned back to Hong Kong when this lab closed).

JS: Stephanie left with Dr Dekker?

EF: Maybe, she was from New Zealand, my friends told me that she went back.

JS: Did Akil Gamal ever do some of the injections and some of the breeding work, and ask you to help him?

EF: Never.

JS: Did you ever see him do injections, or did he ever bring you tissue and ask you to genotype it for him?

EF: No.

JS: You met Dr Gamal once and he called you once to ask you to check on the pups after you left the lab?

EF: Yes.

JS: Do you remember when that was?

EF: Maybe 4 months after, Maybe April and June [of 2010]. (EF then reported, through the Arabic-speaking committee member, that she had not contacted anyone after she realized that the investigation had started, and no one had contacted her either.)

JS: So no one has contacted you?

EF: I don't know, Akil just called me about the pups and Dr Dekker never called me.

JS: We know when you stopped working for Dr Dekker, and you mentioned that Dr Gamal called you this past spring. Was there any time when you stopped working for Dr Dekker before you started working in urology? Was there any break or did your immediately switch from Dr Dekker to the urology lab?

EF: No, there was no gap.

JS: Since you stopped working with Dr Dekker, have you had any reason to go into his former colony room?

EF: I first needed to wait for new pups, then after I did the genotyping for the urology mouse colony, in the same room as Dr Taylor's mice (the former Dekker colony). Everyday I checked the mice for pups.

JS: When did you start working with urology lab mice?

EF: I don't know. (EF then reported, through the Arabic-speaking committee member, that Ms Fathi had left Dr Dekker's lab in July and started going into the animal rooms again in September.)

JS: Do you remember Dr Taylor coming to visit you, and to tell you that she was done with the Dekker lab animals and to not do any more work with them?

EF: Yes. Dr Taylor told me that I did not need to touch the mice.

JS: When you did genotyping, then, who did you give the results to?

EF: Dr Taylor and Dr Gamal.

(At this point, EF then reported, through the Arabic-speaking committee member, that she did remember calling back Dr Gamal in response to his request to check the mice—and she started to become appear more anxious at this point in the interview.)

JS: So you called him back?

EF: I called him once or twice to let him know that the mice pups were born.

JS: Did he ask you to do anything to the mice, or just to call him and let him know?

EF: I just called him to let him know.

JS: Did you do that this Spring, or in the Fall?

EF: I forgot, I don't know, maybe around May.

JS: Of this year?

EF: Yes, maybe I am not right, it seems like a long time...I forget.

JS: And other than letting him know that there were mice there, you didn't do anything to the animals?

EF: No, I just told Akil if the mice were OK or not good.

JS: Did you separate the pups from the mothers or anything?

EF: No, I did not do that.

JS: As you think about this, and you remember anything else, please call me and talk to me about this. It is important that you not call Dr Gamal or Dr Dekker about our wanting to talk to you about this. We need to contact him more officially, and so I am asking specifically that you have no further contact with either of them.

EF: I won't talk to them. Thank you.

In addition, the subcommittee interviewed Mr Marc Waters, a senior research assistant in the urology lab that now employs Ms Fathi (JS = Jim Stevens; MW = Marc Waters):

JS: What kind of interaction did you have with Dr Dekker's lab group?

MW: I pretty much set them up in their laboratory. When their lab was established I helped them get their safety stuff in order, and helped them get set up. If someone was not around and they had students, I would help them with what they were doing and answer questions. Things like that. We worked side by side. I also helped with ordering process. There was Akil Gamal and another woman...a Ph.D. student, I think. She graduated and went back to California or somewhere. Then Akil was left.

JS: We talked before about another woman. Do you recall anybody else in the last year or two?

MW: Yes, Eshe [Fathi] worked for them. I like Eshe. I think that she is a very nice person. But she is nervous. She was very afraid of Stephanie and Dr Dekker; they did not treat her very well. I do not think that there was any dishonesty. I just think that she is a fearful person and I felt bad for her the way she was treated. Dekker expected his staff to do whatever he said to do. His expectation was that you would do it, no matter what. So he kind of ruled with fear and I observed it. If Akil did anything wrong, it was probably under the direction of Dr Dekker, and he probably felt he had no choice.

JS: Do you know of any other examples, that you saw, of Dr Dekker bossing or controlling Akil?

MW: When they had a problem, with a surgery that went poorly, I talked to Akil and said that he needed to let Dr Dekker know, but he was very fearful. He did not want to lose his position and his ability to remain in the U.S.

JS: He was a student at that time?

MW: Yes, he was kind of over a barrel.

JS: The surgery that went badly…do you know if there was a resolution?

MW: All I can remember is that Dr Dekker was visibly angry, and he berated both Akil and Eshe—telling them that they were incompetent. My office was right across from their lab, and I unfortunately could hear everything. He was a lousy boss and I know that Akil was under huge pressure, and I am sure that everyone in the lab felt it, including Eshe.

JS: We asked to interview you because you told me that you had seen Dr Akil Gamal, and Ms Eshe Fathi, together in the animal facility, several months after the Dekker lab closed and Drs. Dekker and Gamal left the institution. Is this correct? And, was this before or after your learned that we were concerned about this?

MW: It was definitely before. It was springtime, and it was very late in a day. I don't think anyone else was around except for Akil and Eshe, but I was still in my office across the hall. I mentioned it to the PI in my lab, and he had seen Dr Gamal too.

Finally, the committee met with Dr Taylor to hear her account of the events. Selected portions from the interview, held on September 19, 2010, follow:

JS: What was your relationship with Dr Thomas Dekker, Dr Akil Gamal and Ms Eshe Fathi?

LT: I have worked with Dr Dekker for many years, as colleagues. When Tom decided to take a new position out-of-state, he asked me to help him by temporarily maintaining his transgenic mouse colony—as a back-up—while he re-established the same genetic line at his new institution. With perfect hindsight, I now think that this was a mistake. Tom should have been required to ship breeding pairs to a commercial breeder for safekeeping. But our agreement was that I would simply maintain the colony and that no more work would be done with them here. When I found that there was clear evidence of new experimentation going on with those animals, I could only assume that these experiments could not possibly have been approved by the animal welfare committee (IACUC) and I had no idea who was doing this. I first went to speak with Eshe Fathi, the research assistant who had been maintaining the colony under Dr Dekker and Dr Gamal. Eshe told me that she was unaware of any new work with the animals, but I nonetheless reminded her not to work with the animals any longer. I am sure she understood me.

Another week or two went by, and I again found that mice in the colony were being caged together for breeding purposes. I also found a dead animal that had had a tail snip. I immediately separated the breeders and culled the pups. Then, a week later, I again discovered new breeding pairs, and so I assumed that some employee in the facility was still breeding the mice—but I had no idea why.

In late June, I received a call from a technician in the animal facility, to tell me that another animal in the colony had died overnight, and that there was evidence at the cage that an injection had been given. Of course these animals should not have had any injections at all, and when I went down to the colony room, it was obvious to me that about 10 mice had had tail snips, and there were cage cards for them that looked new. The cards for these animals were labeled "Akil". I am absolutely certain that none of my staff did this. I then called Dr Dekker about this matter, and he sounded surprised that Dr Gamal had not spoken with me in advance to discuss further use of the colony we were maintaining. I then called Dr Gamal to ask for an explanation, and he initially admitted to me by telephone that he had been in the facility on at least five separate evenings— after work—to complete a quick study with Ms Fathi's assistance. I told him that I needed to report this to the IACUC, and then he told me that I must have misunderstood him…and in the same conversation he actually denied what he had just said! Unfortunately, I never had anything in writing from Dr Gamal…so maybe it is my word against his.

I immediately told you about this, since you chair the IACUC. It was also the same time that I found that five of the mice in my own colony, in the same room as the Dekker colony, were completely missing, and a few others had tail snips. These are expensive animals that I paid for off my R01 grant, for transplant experiments.

JS: What did Dr Dekker admit to you about this matter? What did Dr Dekker say when you told him that someone was doing something to your animals?

LT: The one time I called him about this, to let him know that someone was coming in and working with the animals, Tom said "Didn't Akil call you?" I don't think that Tom ever denied that Dr Gamal came down to work with the animals, but he also never admitted to instructing Dr Gamal to initiate new work in the colony. He did ask, "Isn't Akil still on the protocol?" I reminded Tom that both he and Akil had, of course, been removed from the protocol when they left our employment. We ended that telephone call with Tom promising me that Akil would stop any further work on our campus.

JS: Why do you think that Dr Gamal genotyped the mice (as indicated by tail snips)?

LT: Well, Tom told me that just before leaving the institution, they had submitted a manuscript to *Cartilage*. The peer reviewers for the paper evidently had requested two additional experiments. I think that Akil came down to our hospital, at night, to complete those experiments— but he never bothered to talk to me or ask me for assistance or permission.

JS: Two months after they left, an amendment to one of your other protocols was submitted, requesting a few months of temporary privileges for Dr Gamal to conduct a few experiments on mice in your own colony. Did Dr Gamal ever come down to do that work?

LT: No, Dr Gamal never came down to do the experiments requested in that particular amendment. I did those experiments myself, as authorized by the IACUC.

JS: Did you ever ask anyone how Dr Gamal got [back] into our animal facility?

LT: No, but I have heard from staff in the facility that when people leave there is no enforced policy on return of keys and badges.

JS: Did anyone else that you know of say that they definitely saw Dr Gamal in the facility after his employment ended?

LT: Yes, when I first asked Eshe, she said that she had been with Dr Gamal in the colony room to "check the health of the colony".

Q: In your single conversation with Dr Gamal, did he mention any communications with Ms Fathi?

A: Initially yes, but by the end of the telephone call he had essentially reversed his entire story and denied ever being there.

Q: We have an e-mail from Dr Dekker to the research administrator for the Orthopedics Department, stating that Dr Gamal would not "access the mice", and that Ms Fathi was "overseeing the remaining breeding of mice." Were you aware that Ms Fathi was breeding these mice?

A: She certainly should not have been. I was the PI on that protocol when Tom left, and Eshe had moved to a new lab and should have had no further need to work with the Dekker colony.

Q: Has Dr Dekker reimbursed you for the loss of your own animals, paid through an NIH grant?

A: No, but I have not asked yet.

Q: Did Ms Fathi ever work for you?

A: No, she was already working for urology when I took responsibility for Tom's colony.

Q: Did any of your staff ever work with the animals that were subsequently found to have been genotyped and injected?

A: No, my technician helped to separate the breeding pairs with me, and another assistant helped to add my name onto the cage cards after I became the PI for the project. These two individuals were accompanied by me and did not return to the animal room after these tasks were completed. I think that I was the only one from my lab to check on the animals (other than the animal facility staff), and I checked them infrequently, about every two weeks. It never occurred to me to check them more frequently, because nothing was supposed to be done with them. I guess I have a question for you now, or maybe for the institution in a more general sense. Why is it that ID badges are not electronically deactivated when employees leave the hospital, and since we still use regular keys to get into the animal facility, why aren't these collected intentionally when employees are terminated?

## 5. SYNTHESIS AND FORMULATION

Based on the interviews, handwriting evidence, and e-mail correspondence from the time of the incident, the committee could confidently reconstruct a likely timeline of events leading up to the incident report. The subcommittee's primary findings are summarized below:

1. Dr Taylor's animals, from her own colony, were used without her permission or knowledge. Dr Taylor ordered special mice from Jackson labs, funded by a Federal grant. It appears from both new, handwritten cage

cards, and evidence of both injections and tail snips in both the Taylor and Dekker mouse colonies, at the same time, and in the same room, that her mice were used by Dr Gamal. Notations on Dr Taylor's cage cards indicate that they were injected with Poly I:C which is consistent with the work that Dr Gamal and Dr Dekker were doing, and not part of Dr Taylor's own research. Dr Taylor's special mice were housed in the same room with the colony that previously belonged to Dr Dekker and arrived in the facility on the same date as when a Dekker transgenic mouse litter was born. Verbal communication between Dr Taylor and Dr Dekker around the time when this incident was discovered indicates that Dr Dekker submitted a manuscript to *Cartilage* for publication and the reviewers requested additional experiments in support of the manuscript. It appears that Dr Gamal came down to this hospital in the evenings, to perform those experiments, because the colony with the same genetic strain at his new institution had not yet been sufficiently well established. It is possible that Dr Gamal may have pulled out the wrong cages (in the Taylor colony) to inject with Poly I:C; however it is important to note that at this time, Dr Gamal was no longer authorized to enter the animal facility or to perform procedures on animals at this institution.

2. Dr Akil Gamal's handwriting was on several cage cards of mice that belonged to Dr Taylor. This has been verified by several individuals who are familiar with his handwriting, and also by comparison with known samples of his handwriting on record in the animal care facility.

3. Dr Akil Gamal, after he had terminated employment in July of 2007, asked Eshe Fathi, a former lab assistant in the Dekker laboratory, but who is presently employed in urology, to check on the status of the breeding colony and to call him when new litters were born. After several conversations with the chair of the IACUC, Ms Fathi did admit to doing so on several occasions during the spring of 2010, but she could not recall specific dates.

4. Animals in the colony, formally owned by Dr Dekker and now under the control of Dr Taylor, had their tails snipped and were genotyped during the spring of 2010. Ms Fathi and all other former Dekker lab personnel have denied touching these mice.

5. Dr Taylor separated the breeder mice on at least two occasions during 2010 to stop the growth of the colony, but despite these efforts, breeding continued as evidenced by new litters that continued to be born.

## 6. RESOLUTION

The IACUC subcommittee, charged by the RIO to conduct this investigation, released its final decision on December 4, 2010. In its report, the subcommittee addressed a number of issues highlighted by the investigation and recommended a number of corrective actions to be taken.

1. *Security at the Animal Care Facility:* The security system at the facility was notably "sub-optimal in comparison to current standards at other academic medical centers in this region, and more broadly. First, the facility was accessible via standard metal key-and-lock, and there was no enforced policy for the return of keys whenever employees are terminated for any reason. Secondly, there are no barriers to access, by current employees, to animals maintained by other laboratories for which they have no responsibility. The subcommittee recommended the installation of a modern badge-access electronic entry system, to automatically track entry and egress from the facility 24 h/day, as well as to easily terminate access when an individual is discharged from employment.

2. *Changes in IACUC Policy and Training Procedures:* As a result of this incident, the institution immediately instated a new holding protocol, intended for use when transferring surplus animals between investigators and in future for temporary holding, as would have been appropriate in Dr Dekker's case. In addition, the IACUC chairperson was charged with developing and distributing training materials regarding procedures for reporting animal use violations to the IACUC. These training materials have also addressed a notable problem identified during the course of this investigation, namely, that there was a fairly lengthy delay in the reporting of the incident to the IACUC, as Dr Taylor and her staff attempted to investigate the matter on their own first, likely resulting in a loss of information pertinent to this case.

3. *Accountability and Reporting:* The IACUC investigative subcommittee agreed that the internal investigation into these events recommended that the incident, and the institution's response, be reported to the NIH as well as to appropriate authorities (the RIO) at the medical center that now employed Drs Dekker and Gamal. In their final report, the committee chair noted that he had tried to speak with Drs Dekker and Gamal by telephone to reach a full understanding of the events, but he was unable to secure their cooperation.

Because the main persons likely involved in the misuse of animals (Drs Dekker and Gamal) were no longer employed by this institution, at the time of investigation, the subcommittee took no action against any individual for misconduct during the events in question, relying on the assumption that the corresponding authorities at their new institution are now responsible for the conduct of their research faculty. Although the subcommittee did not explicitly note this in its final report, the procedure for reporting and investigating animal use violations by nonemployees is unclear, with no certainty over which entity has jurisdiction over the incident. Most certainly, it is the local institution's responsibility to ensure that funds from a Federal granting agency are only used for the purposes designated in any given funded grant. But who is responsible when the misuse occurs by nonemployees who have trespassed into a facility without permission? Their current employer also had no knowledge of these events, and no way of policing their activities either. Finally, although this institution could not take any specific action against Drs Dekker or Gamal, there is at least one current employee—Ms Eshe Fathi—who did provide some support to Dr Gamal by checking on the colony for him, and it became increasingly clear over the course of this investigation that she understood that this was not appropriate or in keeping with institutional and IACUC policies.

## 7. COMMENTARY

At first glance, the facts uncovered by the IACUC subcommittee investigation make this incident seem like a cut-and-dry case. A postdoctoral fellow, aided by a lab technician still employed at the institution, continued to use animals and facilities from a past project at a previous place of employment, and all without permission or IACUC approval. In addition, he made use of animals intended for another project without authorization. To fully understand the ramifications of this investigation, it is important to identify and examine the areas in which unanswered questions were raised: international culture and gender relations in the American academic workplace, jurisdiction over internal and multi-institution incidents, legal aspects of academic and research violations, and individual responsibility in a multiperson lab settings.

### 7.1 International Culture and Gender in the American Academic Workplace

One of the interesting issues that the IACUC subcommittee encountered while investigating and adjudicating this case was the situation of Eshe Fathi, the lab technician formerly of Dr Dekker's lab who had been transferred to another lab within the hospital, on Dekker's departure. The fact that Ms Fathi, an Egyptian national, was not fully comfortable with English as a second language, nor with American academic culture and societal norms, became clear to the committee over the course of the investigation. Ms Fathi was allowed to submit a narrative report to the subcommittee in her native language, and a member of the committee who is a native Arabic speaker was present for all of her interviews. Additionally, as a foreign national woman in this academic workplace, Ms Fathi was in a minority and working under the direct auspices of an Egyptian male doctor (Dr Gamal), both hailing from a culture with strictly defined and enforced gender roles in workplace environments. Although workplace sex discrimination is more-or-less ubiquitous worldwide, the United Nations Educational, Scientific and Cultural Organization (UNESCO) lists Egypt as a country with one of the worst records in this regard. Ms Fathi is a recent emigrant to the United States, from a country in which women are simply expected to heed the instructions of male counterparts in positions of authority, combined with the apparent authoritarian style with which Dr Dekker managed his staff, both suggest that the manner in which Ms Fathi responded to requests from her superiors in the lab was influenced by a number of cultural and social factors that add an important dimension to Ms Fathi's role in the case.

In a related concern, the IACUC subcommittee also had to decide whether Akil Gamal knowingly misused the animals at this institution, or if his actions were motivated from an unintentional misunderstanding of the terms of his separation from the institution to take a new job elsewhere. While Dr Gamal did not appear from the interviews to be involved in creating the bullying overtones in the Dekker lab, he was working with the same person (Dr Dekker) who "expected you to do whatever he said to do...he kind of ruled with fear..." (Marc Waters interview, above). While this is only one possible explanation for Dr Gamal's motivation to conduct unauthorized research in the facility, it is clear that the interpersonal style of the principal investigator could have affected judgment and decision-making by his subordinates.

Nonetheless, Dr Gamal provided conflicting stories (within a single telephone call with Dr Taylor) and he refused to be interviewed at all by the committee chair. It seems clear that he was aware that he was breaking institutional and IACUC policies by entering the facility in evenings to conduct experiments on animals without proper authorization by the animal welfare committee. Dr Gamal, also an Egyptian national, may have felt pressured by the high-paced environment of the American research hospital, and a need to please and respond to a demanding mentor, without fully understanding both the responsibilities and limitations of his position as a postdoctoral fellow. This case raises questions about whether this institution provides enough of a solid introduction and acculturation, to U.S. standards and ethical codes of conduct in academic and laboratory settings, for trainees and faculty who are foreign nationals.

## 7.2 When Should a Research Integrity Breach Investigation be Relegated to State Authorities?

If the setting in question were not a research laboratory that abides by both National Institutes of Health (NIH) Office of Research Integrity (see Appendix 1) and its own policies that govern research conduct (see example, Appendix 2), the triage of this case might have been clearer. A report could have been filed with State Police (as Dr Gamal likely crossed state lines, given the location of his current home and employer), and he could have been the target of a police investigation for criminal trespass, as well as theft (use of Dr Taylor's mice without permission). Moreover, what is the institutional liability, given that a former employee had such easy access to an animal facility with substandard security in place at the time?

Although some of the decisions and actions, by both Dr Gamal and Ms Fathi, may have been at least partly influenced by cultural factors and a possible punitive interpersonal management style by their principal investigator, these factors would likely have had little bearing on how this case might have been managed in a court setting. Should academic violations, that could be possibly construed as legal infractions, be reported to local police authorities? When should an internal institutional research conduct incident merit civil or criminal court involvement, and how should such a decision best be made by the RIO?

In this case, because the misuse of animals ultimately resulted in a loss of NIH grant money for the project for which it was earmarked, the internal affair also required Federal reporting to the NIH, which could then theoretically take further action if needed. The NIH has established clear guidelines for compliance and the reporting of violations, but there is a noticeable lack of information regarding follow-ups to investigations and accountability for violations. This criticism has not gone unnoticed, and the NIH is currently in the process of expanding and improving its "education efforts" regarding "compliance with Federal Conflict of Interest (FCOI) regulatory regulations."[1]

[1] Website: http://grants.nih.gov/grants/guide/notice-files/NOT-OD-12-159.html.

This effort indicates a growing concern with the oversight and resolution of compliance issues, which presumably would cover a myriad of types of violations with respect to institutional management of grants received.

## 7.3 Multi-institutional Involvement in Research Integrity Cases

A third question raised by the investigation was that of jurisdiction over the parties involved in the case. Because Drs Dekker and Gamal were no longer employed locally at the time of the incident, the internal IACUC investigation could not compel them to provide statements nor could it enforce any correctional action it recommended regarding the former employees. In its final resolutions, the committee stated that the internal investigation had "proceeded as far as it can go" and relegated the task of further investigation of Drs Dekker and Gamal's involvement to the RIO at their current academic medical center. The RIO at that institution (in a conversation with his counterpart locally) agreed that this case appeared to be a serious matter, and he agreed to issue a "stern warning" to Dr Gamal about any such future conduct, but this is the limit of his own abilities to pursue the matter, since these events all transpired outside of his institution. How could either institution manage this case effectively?

## 7.4 Individual Responsibility in a Lab Setting

Ultimately, who is responsible for the misuse of animals that occurred at this institution? Or, put in terms more specific to this case, should Dr Dekker be held accountable for the actions of his postdoctoral fellow and (former) research assistant? Recently, the prestigious journal *Nature Chemical Biology* adapted its authorship rules to stress that the "senior researcher from each group assumes responsibility for the team's contributions to the project" and ensures that "each senior scientist of a multi-group collaboration accepts responsibility for the primary data generated by his or her group" (Editorial Board, 2009). If we accept these standards as reasonable, it makes sense that a senior investigator should be held responsible for the research integrity of his entire group as well, since, presumably, the entirety of the research falls under his or her purview.

Although the roles played by the PI in this case were not in the forefront of the investigation, the actions of the researchers within the lab point to some inadequate amount of oversight over the research team in the Dekker and urology labs. Dr Dekker was apparently unaware of the specific actions that were being taken with the colony still left behind, and he was unable or unwilling to definitively clarify whether Eshe Fathi or Akil Gamal were working with his permission on additional experiments.

This case highlights a new trend in academia to explicitly define the responsibilities of a lab's principal investigator, to hold such individuals accountable for the actions of their staff and trainees. Dr Dekker, and the head of the neighboring urology lab, if not intentionally neglectful of their responsibilities, perhaps felt too busy to oversee the daily work taking place under their auspices. In the push to secure increasingly scarce grant funding, which requires (in part) more peer-reviewed publications as marks of academic success and future potential, perhaps some of the core integrity of the research itself might be jeopardized.

## QUESTIONS FOR DISCUSSION

*Parties*: Ms Eshe Fathi, defendant, hospital lab technician; Dr Lorinda Taylor, responsible for colony, submitted incident report; Dr Thomas Dekker, project lead; Dr Akil Gamal, postdoctoral fellow of Dr Dekker; Dr Jim Stevens, Chairperson of the IACUC, led inquiry of complaint.

1.  This case centers on an unlawful break-in to an animal facility, and the inappropriate and unapproved use of transgenic mice. This illegal and ethical breach is not covered under the NIH Office of Research Integrity (ORI) definition of research misconduct (see Appendix 1). Moreover, mice are not covered as a reportable species to the NIH Office of Laboratory Animal Welfare (OLAW). Is the ORI definition of research misconduct sufficiently broad? Should individual institutions modify this definition in their own policies? If so, how should research misconduct be redefined?

2.  Although the mice used in this case are not an OLAW-covered species, these are nonetheless expensive animals, paid for by a Federal grant from an agency whose budget is supported by public tax revenue. Does this fact influence your thinking about this case and how it was resolved? If you were the RIO for this case, would you have reported this to state police for investigation? If so, why?

3.  The lab assistant, Ms Eshe Fathi, has only been working in the United States for a very short period of time, and she comes from a country with a terrible record for sex discrimination, with women expected to show obedience to men in any public setting. Dr Gamal comes from this same country and society. To what extent may have Ms Fathi's actions been influenced by her self-expectation to heed instructions from a male doctor? Should these cultural issues be factored in, as the RIO determines Ms Fathi's role and continued employment as a staff member? If so, how?

4. The animal facility was secured with traditional key locks, and there was no enforced policy in place to ensure that ID cards and keys were returned when a staff member left employment at the institution. What role did the institution have in this case?

5. How can RIOs avoid the problem of delayed reporting of suspected misconduct? Won't institutional employees often hesitate to accuse their colleagues of serious misconduct?

6. Why was Ms Fathi, the only current employee involved in the misconduct, not sanctioned? Did Dr Dekker's authoritative manner and the impression that Ms Fathi was just following his orders play a role?

7. What is the best way to sanction misconduct by former employees? Reporting to ORI? Indeed, Federal funds were implicated here.

8. Should the RIO have simultaneously commenced an institutional investigation and reported the suspected trespass to state authorities?

9. Could Dr Taylor have brought a civil suit against Drs Dekker and Gamal for the misappropriation of her property?

## REFERENCE

Editorial Board, 2009. Assigning responsibility and credit. In: New Policies Refine the Responsibilities of Authors and Require Author Contribution Statements in Nature Chemical Biology Research Papers. Nature Chemical Biology, vol. 5 (10), pp. 697.

## CHAPTER 5

# Research Assistants Coming Forward with Concerns about Perceived Behavior of Principal Investigator

## 1. PRESENTING COMPLAINT

An undergraduate student working for a summer in a clinical research setting for individuals with diabetes came to the director of the clinic expressing concerns about instructions from the principal investigator, Dr Cosgrove, on a study on behavioral treatments for adults with diabetes who were also obese. The goal of the study was to provide more effective and healthy weight loss strategies to these individuals that would at the same time improve their management of their diabetes. Subjects were eligible for the clinical trial based on a combination of a history of diabetes managed with insulin but who had failed previous weight loss programs. Patients regularly used and reported the results of their home blood sugar monitoring which was used by the clinical team as data regarding their compliance with their diabetic treatment. Hemoglobin A1C values were also checked quarterly as a measure of blood sugar control and were used to adjust insulin and other drug management. Patients were eligible for the trial if they had failed at least one previous weight loss intervention and if their home blood sugar monitoring over the past month coupled with their most recent hemoglobin A1C value (assayed quarterly) indicated effective blood sugar management over the past 3 months. The student alleged Dr Cosgrove had instructed him to change/adjust results of the patient's most recent hemoglobin A1C values in the database such that patients who appeared less compliant with their diabetic treatment based on a borderline high hemoglobin A1C were nonetheless eligible for the behavioral treatment trial targeting weight loss.

*The Management of Scientific Integrity within Academic Medical Centers*
http://dx.doi.org/10.1016/B978-0-12-405198-0.00005-9

## 2. BACKGROUND AND HISTORY

Dr Cosgrove was a young clinical investigator from internal medicine and endocrinology with extensive experience in the clinical care of patients with difficult-to-manage diabetes and related chronic health problems including obesity. Especially focused on weight loss and lifestyle change as particularly challenging for many adults with diabetes, Dr Cosgrove had designed a number of behavioral interventions focusing on weight loss and reducing the debilitating consequences of obesity and diabetes. The clinical problem was especially challenging given the high rate of obesity among many adults with diabetes and conversely how obesity complicated diabetic control in these patients. Dr Cosgrove had just received an NIH grant to support a randomized clinical trial, and she was just beginning enrollment of patients in ongoing weight loss interventions.

The student working in the clinic was an undergraduate biology major and a rising senior with an interest in medicine. His work in the clinic was as a summer work study program through which he would also complete his senior thesis. He worked on several projects in addition to the one with Dr Cosgrove and was familiar with protocols regarding screening of subjects with chronic illnesses for treatment studies. He had worked in other clinical research studies throughout his college years. He was responsible for direct interactions with the potential research subjects as well as the treating clinicians and also for data entry and maintaining databases across several clinical trials in the clinic. He interacted regularly with Dr Cosgrove around the beginning of enrollment for the treatment study and understood the study was in the initial pilot/feasibility phase. He reported that Dr Cosgrove told him to consider patients with borderline high hemoglobin A1C values as "normal" and interpreted the communication to mean he should alter the data in the study database which he did. However, he did not alter the reports from the laboratory though he noted on the reports that results were considered "in the normal range" per Dr Cosgrove's instructions. He also understood that because the human subjects protocol for study eligibility required subjects to have both a clinical history of compliance with diabetic management as indicated by their home blood sugar monitoring and a normal hemoglobin A1C, Dr Cosgrove's instructions to him to alter the subjects' data was intentional to make the subjects eligible for the treatment trial by the criteria of the extant human subjects protocol.

## 3. RELEVANT EVIDENCE

Once the student came forward with his concern and sought the counsel of the clinic director, the RIO first sought the advice of two senior faculty members familiar with clinical trials about whether or not these allegations, if true, would constitute potential scientific misconduct. The two faculty members were concerned both about the alleged instructions to change the clinical database and the resulting impact on clinical care, that is, clinicians would see their patients' with "normal" hemoglobin A1C values as indicating reasonable diabetic control over the past 3 months. The faculty members were also concerned about a potential violation of the human subjects protocol by altering data so as to alter eligibility for the treatment trial. Thus, the faculty members recommended to the RIO to move forward with a more detailed investigation of the allegations and at the same time, to consider when and how to involve the human investigation review committee.

To obtain additional evidence, the RIO met with the student prior to notifying Dr Cosgrove of the allegation and obtained copies of the clinical research files for the subjects enrolled to date, their original reports from the laboratory, and the electronic database in which the clinical and research data were recorded. The treating clinicians had access to the data showing the hemoglobin A1C values (as well as other clinical data relevant to the patient's overall health). Once these data were secured, Dr Cosgrove was notified of the allegation.

## 4. CASE MANAGEMENT AND ADJUDICATION

The faculty committee conducting the investigation interviewed Dr Cosgrove, the student, another colleague whom the student consulted, and the director of the clinic. In Dr Cosgrove's interview with the committee, she explained that the treatment study was in the very early phases and she was trying to understand if patients would find the weigh loss approach acceptable and feasible. She was not using the hemoglobin A1C values to monitor treatment progress but only as one component of study eligibility. Indeed, for her study, weight loss was the most relevant outcome. Dr Cosgrove went on to explain that after submitting the human subjects protocol, she had realized that her criteria for a hemoglobin A1C value in the normal range was probably too stringent, especially for patients struggling with both obesity and diabetes. Her criteria, while clear from a research eligibility point of view, simply did not meet clinical reality in patients with such a complicated medical profile. In other words, such criteria were so strict as to limit the generalizability of her clinical trial. Thus, she had revised the eligibility criteria to allow for patients to be recruited into the treatment study even if their hemoglobin A1C values at the time of screening for the study were borderline high. However, she had neglected to request an amendment from the human investigation committee to alter the eligibility criteria to reflect this particular clinical reality, and she had not been explicit with the research team and especially the student researcher. In her memory, she had told the student to consider the patients with borderline high hemoglobin A1C values but home glucose monitoring from the past month indicating adequate blood sugar management as having "normal" hemoglobin A1C values meaning they were eligible for enrollment into the treatment study. But she had not intended her statement to be interpreted as instructions to actually alter the lab results in the research database.

Dr Cosgrove was surprised and shaken by the allegations. She denied that she had changed the eligibility criteria simply to speed up enrollment into the trial as there was a sufficient flow of patients and especially so for the number of clinicians working in the clinic who could see the patients and provide the behavioral treatment. Nonetheless, Dr Cosgrove did understand that because the study database was available to treating clinicians, the altering of any lab results in the database might impact patient clinical care. She was very concerned both about the impact on patient care as well as on her own clinical research career with an inquiry regarding scientific misconduct. At the time of her interview with the faculty committee, she had already requested an amendment to her human subjects protocol for the treatment protocol and had met with her research team to explain the new eligibility criteria.

In their meeting with the faculty committee, the student, Peter, was forthcoming and thoughtful. He explained that he did not fully understand Dr Cosgrove's statement that he should consider the borderline high hemoglobin A1C values as normal and he had asked another more senior research assistant, Emma, to join him in another conversation with Dr Cosgrove. The faculty committee also met with Emma who as the more experienced research assistant usually took responsibility for overseeing the student researchers. Each stated that they both asked Dr Cosgrove to clarify her statement about considering the subject's lab results as "normal" and that

Peter specifically asked if he should change the entry in the database. Peter stated that Dr Cosgrove had affirmed several times to enter the result in the database as "normal." Emma generally agreed with Peter's account of their meeting together with Dr Cosgrove, but she was less clear in both how many times Peter had asked Dr Cosgrove about what to enter or change in the database and also how clearly affirmative Dr Cosgrove's response had been. Emma allowed that Dr Cosgrove may not have clearly understood their question and did not remember Peter's repeating the question to clarify each of their understanding. Peter took care to note on the laboratory reports that the borderline high results considered "normal" per Dr Cosgrove and in so doing, left an audit trail of the change in the electronic database.

The director of the clinic, Dr Taylor, was concerned if there would need to be action with Peter for altering data in the study database especially since that database was available to the treating clinicians and could have influenced their work with patients. Because it appeared that Peter had acted on Dr Cosgrove's instructions, however unclear, Dr Taylor felt no action against him was needed though she was planning to meet with him and offer mentorship in how to obtain clarification from a more senior colleague before doing something he was uncomfortable with. Dr Taylor stressed that Dr Cosgrove had always been honest and very careful in her work and that this incident seemed out of character and not consistent with her approach to research and clinical care thus far.

The faculty committee concluded that Dr Cosgrove changed the inclusion criteria for her study without informing the human investigation committee and also without speaking to her research team. While there was the possibility of an impact on patient care, in reality, the change had negligible, if any, impact on the patients' treatment. The unaltered laboratory reports were available in the clinic research charts and thus treating clinicians might note the discrepancy with the results in the research database were those questions to arise. The laboratory results were also not used as outcome measures for the treatment trial and were only informing eligibility, albeit inaccurately since a number of patients at the time of screening for the study had had borderline high hemoglobin A1C values over the last several assessments. The committee also noted that Dr Cosgrove had obtained recent approval with no questions from the human investigation committee for the change in the protocol to determine patient eligibility for the treatment trial.

At the same time, the faculty committee was dismayed by Dr Cosgrove's seeming sloppiness and inattention to detail both in revising her human investigation protocol and in conveying changes in eligibility criteria clearly to her research team and this concern carried over to their questions about what Dr Cosgrove really conveyed to Peter. The committee was unable to confirm whether or not Dr Cosgrove instructed Peter to alter the database or simply to consider the patients eligible for the study despite their border-line high lab values. Though the findings were ambiguous, the faculty com-mittee was sufficiently concerned at the possibility that Dr Cosgrove may have instructed Peter to alter the database and that this possibility was a significant departure from responsible research conduct. Thus, the faculty committee recommended further investigation into the matter.

While the faculty committee was reviewing the matter, the institutional human investigation committee also reviewed the protocol lapse, that is, Dr Cosgrove's revising eligibility criteria, without informing the human investigation committee. They determined that no patient harm occurred and that Dr Cosgrove understood her unfortunate and unacceptable lapse in attention to the proper consent and oversight for human subjects research. Dr Cosrove's requested amendment regarding the eligibility cri-teria was approved. Also, while the faculty committee was at work, the RIO consulted the NIH Office of Research Integrity (ORI) because of Dr Cosgrove's NIH funding for her treatment trial. After presenting all of the available evidence and analysis, officials at ORI made the decision not to pursue the matter. Their decision was based on the lack of evidence that Dr Cosgrove acted recklessly or with intent to deceive the scientific com-munity and the altered data were not disseminated in a publication or public presentation. With this response from ORI and also the awareness that first, there was likely nothing more to be discovered in a more formal investigation and second, the faculty committee had thoroughly inter-viewed all relevant witnesses, the RIO discussed with the dean the possi-bility of accepting as final the findings of the faculty committee.

## 5. SYNTHESIS AND FORMULATION

The faculty committee felt that there was an unfortunate cascade of com-munication lapses that led to the circumstances they were asked to review. First, the student, while spending other summers and academic semesters in clinical research settings, was still relatively inexperienced. He understood the research protocol he was asked to implement, the criteria for enrollment in the study, and how to enter data and manage a database. But he was far less experienced in knowing when and how to seek clarification and when to say simply he did not agree with what he thought he was being asked to do. It was laudable that he had asked his more senior research assistant colleague to join him in a meeting with the principal investigator and also courageous that he had sought out the clinic director. But his relative inexperience with communicating with those more senior than he around decisions and instructions came together with Dr Cosgrove's urgency to get her study started, her busy clinical schedule into which she was interweaving research, and her own inexperience in directing a clinical research team. She also had difficulty appreciating that students and young research assistants might not understand the clinical nuance she thought was obvious, that is, subjects with as complex a disease as diabetes combined with obesity might often have lab values indicative of less than adequate blood sugar control over several months. She had simply made her initial inclusion criteria too strict and not consistent with clinical reality. But she did not appreciate that others less experienced would be unable to make that distinction and she also neglected to take the simple step of requesting a protocol revision. Nonetheless, her instructions to the student were at least not clear and at worst directing alteration of the laboratory data to enroll subjects more quickly. Relaxing the eligibility criteria without notifying the human investigation committee or the research team was unacceptably sloppy.

## 6. RESOLUTION

The decision was made to accept the faculty inquiry committee's findings as final without additional investigation. A further decision was made not to stop Dr Cosgrove's study or bar her from clinical research. Rather, the dean decided to require all of Dr Cosgrove's protocols be closely monitored and approved by a senior faculty member for 5 years and that a letter to this regard be maintained in Dr Cosgrove's faculty file. Such a letter would not be available to outside inquiry but only to someone reviewing the file, for example, at the time of promotional consideration or should any additional concerns about her conduct of research arise.

## 7. COMMENTARY

The fundamental issue illustrated by this case is the importance of clear and documented communication between a principal investigator and his or her staff. Even if initially verbal, these communications should be written and sent to all research staff to serve as an archival record for the research team. All too often, principal investigators start their research with no training in managing teams or in ensuring smooth and transparent communication across the team. Indeed, communication is one of the most important factors in building and sustaining productive and creative research teams (Winowiecki et al., 2011) especially when those teams have individuals of different experience and disciplinary backgrounds.

Communication is only made more challenging when students or other team members junior to the principal investigator or individual overseeing the research are uncertain about what they are to do or about the instructions they have been given (or what they perceive they have been asked to do). A critical skill is learning how to ask for clarification in situations where students and staff may feel vulnerable or at the least not sufficiently empowered. These kinds of open communications among research teams depend not only on attention to facilitating discussions but also implementing management approaches that allow all to feel equal and responsible for the team or lab's success. Rarely are directors of research laboratories also trained in basic team management, but learning some basic skills in these areas may facilitate not only more effective communication across different levels of experience and status in the lab but also enhance productivity and deter potential misconduct or at the least sloppiness (HHMI, 2006).

Especially when the complainant is far junior to the laboratory or project director, there is often concern about the consequences of reporting potential scientific misconduct. Regretfully, there is evidence that complainants often suffer adverse consequences when their identity is known. In a 1995 self-report survey (Research Triangle Institute, 1995), over 60% of complainants reported at least one of the following consequences: being pressured to withdraw their allegation, being ostracized by colleagues, suffering a reduction in research support, or being threatened with a lawsuit. Approximately 10% noted significant negative consequences, such as being fired or losing support. However, the majority of those suffering the most severe impact on their careers reported that they would still be willing to come forward with allegations again (Research Triangle Institute, 1995). This potential for adverse consequences for those in training, such as postdocs, graduate students, or undergraduate students, if they come forward with concerns about possible scientific misconduct adds a special institutional responsibility to structure protections and also allow for anonymous allegations.

In many cases, the respondent often suffers adverse consequences even if allegations are not substantiated. In this particular example, the faculty member's ability to continue her work was protected by the process and she was encouraged to continue her studies but under the close mentorship and oversight of a senior faculty member. Still, she was concerned about the impact on her career. No doubt these kinds of allegations are disturbing and often have a prolonged disruptive impact on an individual investigatee even if no evidence of misconduct is substantiated. There are several reasons for such disruption. One is the sheer time such inquires may require that only prolong the individual's worry about the possible outcomes. Another is the potential impact on an investigatee's reputation no matter how careful committees are to protect confidentiality and consider breaches in that confidentiality quite serious. Still even subtle discussions of the process may happen. Other staff may see representatives of the RIO's office in the lab. A staff member meeting with a faculty review committee may ask for advice from a colleague or even speak in general about the process. A third potential disruption for an investigatee because of rare but nonetheless possible breaches of confidentiality is the discouraging of postdoctoral fellows coming to a research program or encouraging staff to find other laboratories to join, each because of perceptions about possible negative consequences for the investigatee in question. In any case, undertaking an inquiry or investigative process is always a serious matter with potentially disruptive effects for an investigatee even if allegations are not substantiated.

## QUESTIONS FOR DISCUSSION

1. How might the research assistant have further clarified Dr Cosgrove's instructions? Could further clarification of the use of written instruction have obviated the need for the complaint and investigation?
2. Might there have been more evidence discoverable in an investigation? What would have been the potential advantages of proceeding to a full investigation?
3. Are there other consequences that may have been put in place regarding Dr Cosgrove's actions? Are there any you would recommend?
4. What procedures might Dr Cosgrove implement for her research team that ensure better communication and more checks and balances?
5. The interactions of different regulatory agencies and systems can be complex. Discuss how those intersections influenced this case. Do you think the NIH/ORI acted correctly in not pursuing the matter?

**6.** How long did Peter proceed in altering the database entries before bringing his concerns to the clinical director and does this consideration alter his culpability in altering the database? Did he have a responsibility to come forward immediately based on his knowledge that research data were being potentially falsified in a manner that could affect patient treatment?

**7.** This case raises the potential for problems resulting from using undergraduate researchers. Should undergraduate researchers be provided any mandatory training on scientific integrity before they are allowed to assist in the lab? Is any training provided to researchers employing undergraduate students in connection with studies? Perhaps those in Dr Cosgrove's position should be reminded of the fact that undergraduate students may require more detailed explanations than experienced researchers.

**8.** Did the ORI's decision to not pursue the case have any influence on the RIO's ultimate decision not to proceed with further investigation?

**9.** How should an institutional inquiry committee evaluate witness credibility? In this case, Dr Cosgrove and Peter told significantly different stories about the former's instructions. Should Peter's version have been given more weight since he arguably had less personal investment in the outcome?

## REFERENCES

Howard Hughes Medical Institute, 2006. Making the Right Moves. Howard Hughes Medical Institute, Maryland.

Research Triangle Institute, 1995. Consequences of Whistleblowing for the Whistleblower in Misconduct in Science Cases. Report submitted to Office of Research Integrity. http://ori.dhhs.gov/multimedia/acrobat/final.pdf.

Winowiecki, L., Smukler, S., Shirley, K., Remans, R., Peltier, G., Lothes, E., King, E., Comita, L., Baptista, S., Alkema, L., 2011. Tools for enhancing interdisciplinary communication. Sustain Sci Pract Policy 7, 74–80.

CHAPTER 6

# Questionable Mentorship and Oversight of Federal Grant Funding

## 1. PRESENTING COMPLAINT

Dr Matt Saltzman, the Chief for the Division of Pediatric Research, came to meet with the Research Integrity Officer (RIO) to discuss a personnel matter involving a young faculty member who has been supported on a large NIH center grant. Dr Saltzman has been increasingly concerned that this individual, Dr Tien Wong, has not made any progress on his projects—funded by the center grant—over the past 1.5 years, despite his regular submission of time-and-effort (T&E) cards and reimbursement and salary for services rendered on that grant. The principal investigator (PI) for the center grant, Dr Jack Bartlett, and his advisory board for the center, are now recommending that Dr Wong, as well as his specific project, be eliminated. Dr Saltzman's specific complaint is that Dr Wong has been inappropriately charging T&E to the Federal Grant without completing the work as outlined in the grant application.

## 2. BACKGROUND AND HISTORY

Dr Wong, trained as a protein biochemist and by all accounts did a superb job as a graduate student and as a postdoctoral fellow. He practically lived in the laboratory and chose to work nearly constantly on his research. As a result of his very high productivity, an effort was made to create a junior-level faculty position and to fund his employment through new work on a center grant with several interlaced component projects. The project that was both available and selected for Dr Wong involved the use of live animals and the surgical insertion of a live virus directly into the wall of the small intestine of rodents. Although he had no prior animal laboratory experience, the PI for the center grant assumed that he could learn and master these new procedures.

*The Management of Scientific Integrity within Academic Medical Centers*
http://dx.doi.org/10.1016/B978-0-12-405198-0.00006-0

The center grant was awarded to the institution, Dr Wong was hired into this new faculty position, and he commenced work on the project. He is described as an intensely hard worker, who maintains bizarre hours in the laboratory and very often "sleeps under his desk" rather than going home. He has very profound difficulties with the English language, and although he has struggled to become conversant in English, this has apparently not met with much success.

Dr Wong is described by Dr Saltzman as having virtually no social skills whatsoever, and in fact Dr Saltzman referred to him as an "idiot savant." Examples offered include Dr Wong's admonishing a coworker that she should not have become pregnant because it would slow down the pace of her work in the lab, and he also suggested to a newly hired postdoctoral fellow that she marry him so that she would not have to worry about visa issues and can spend more time working in the lab. Clearly, Dr Wong has issues understanding appropriate social conduct in the workplace, and he has difficulties interacting with colleagues and supervisors within both structured and unstructured settings.

## 3. RELEVANT EVIDENCE

As a junior faculty member in this new center, Dr Wong was assigned to Dr Brian Forrester as his primary advisor or mentor. It was Dr Forrester's responsibility to oversee the quality of Dr Wong's work and to monitor his career development. Dr Wong's project was launched in January, 2009. Sixteen months later, in April of 2010, Dr Forrester informed the center's governance committee that Dr Wong had made little, if any, progress toward the central aims of his project. Rather, he had continued to work on the questions and with the laboratory techniques that he mastered during his recently completed postdoctoral fellowship. Then, in July of 2010, Dr Wong presented his current research to the center's "data club," and he focused entirely on his prior work despite his advisor's request that he both complete an experiment for the core project and also present those same data to the group. Following this seminar presentation, both Dr Forrester and Dr Bartlett met with Dr Wong to express their disappointment and concerns over his continued lack of scientific progress on his core project. Later that month, the leadership of the center held a meeting to determine how best to proceed in correcting this situation, and Dr Forrester reported at this meeting that Dr Wong had expressed significant distress that he was not comfortable conducting small animal surgery, and that he probably should never have been "inserted" into the center's grant application in the first place.

By mid-August of 2010, the center's leadership committee decided to sever funding on Dr Wong's project, and one week later the Institutional Animal Care and Use Committee notified Dr Wong that his project had been terminated (and all of his research activities were to be immediately halted) due to failure to submit appropriate progress reports. In September of this year, Dr Saltzman requested his meeting with the RIO to report that Dr Wong had—over the past 15 months—made inappropriate use of Federal grant funding by knowingly falsifying his T&E reports and charging his salary to the center's grant without completing the expected work on his project.

# 4. CASE MANAGEMENT AND ADJUDICATION

On review of Dr Wong's record of productivity as a recently graduated postdoctoral fellow and as a laboratory scientist, the RIO determined that Dr Wong was actually quite productive. He successfully applied for and received a competitive N.I.H. R03 grant (as principal investigator), and he had published (or had *in press* at the time of this review) several high-profile articles in such prestigious journals as the *Proceedings of the National Academy of Sciences* and the *Journal of Biological Chemistry*. In his response to the complaints raised by Dr Saltzman and Dr Bartlett, Dr Wong provided a difficult to read and date-dense report, describing his personal progress on the grant-funded work in question. On review of this report, it was clear to the RIO that he had made some measurable progress on two of the four aims listed in his section, but not on two other aims of the grant. Importantly, these two latter aims required the use of live animals for which Dr Wong had no prior experience. In a conversation about this report, Dr Saltzman voluntarily stated that Dr Wong "probably should not have agreed to address those anyway, given his range of skills." This raised two questions for consideration: (1) Why might Dr Wong have committed to complete a work product for which he not only had no prior experience, but also clearly felt uncomfortable acquiring the required skills (i.e., live animal surgery), and (2) Why would the principal investigator for this center grant even offer such a position to a new faculty member who lacked some of the required skills (for the specific project and its aims) in the first place?

With regard to the first question, perhaps Dr Wong allowed himself to be convinced to assume his role and responsibilities on the grant because he had substantial difficulties with English language comprehension. He had a marked tendency to nod yes and/or say "yes" to requests, in order to please his mentors/supervisors, even if he did not fully understand particular instructions or questions. Alternatively, he might have been eager to accept the position as a new faculty member, funded through this grant, because it would allow for a work visa to allow him to remain in the United States and eventually apply for permanent resident status. Finally, even for citizens of the United States, new academic positions in medical centers are competitive, and this offer probably seemed to Dr Wong as one he could hardly afford to decline. More than likely, two or all three of these potential reasons factored in to his decision to accept his position in the center.

The second question, raised by the RIO, centered on why a seasoned, experienced senior investigator would assign an individual, without all the proper qualifications, to sections of a large center grant in the first place? The award of an NIH center grant to any institution is hugely competitive, and the preparation of the actual grant application is a massive undertaking. All details must be provided with great care and thoroughness, and there is little room for speculation or promises. Hence, listing "to-be-determined" for key scientific positions is usually avoided at all cost. These grant submissions are also subject to strict timelines for preparation, and equally strict deadlines for submission. At the time that this grant application was being made, Dr Wong was a postdoctoral fellow in one of the laboratories that was central to the application, and he most certainly did possess at least some of the required skills (after all, he did make reasonable progress on two of the four aims assigned to him). The RIO suspected that his selection was a matter of convenience, and he was thus fit in to this application like a necessary piece of a larger puzzle. This decision, for the principal investigator, likely seemed like an efficient decision, and a hiring issue that could be checked off his long list of things to do as part of "good grantsmanship" in preparing a competitive application.

In managing this case, the responsibilities of the RIO center on managing the process of determining the likelihood of research misconduct by Dr Wong. Additionally, if there was clear evidence of intentional misconduct by his supervisors, then the RIO could pursue this as well. Unless an employee is to be terminated as a result of clear research misconduct, the RIO does not directly manage the personnel decisions made by chiefs of various centers or departments. In this case, the director of the research center, on consultation with his senior leadership committee, made the decision to terminate Dr Wong's activity on the center grant and, hence, the funding for his faculty position. He left the institution toward the end of 2010, to return to his home country.

## 5. SYNTHESIS AND FORMULATION

The RIO rapidly decided that there was insufficient reason to suspect any breach of research misconduct by Dr Wong, and so no initial inquiry committee was convened for any further exploration of the specifics of Dr Saltzman's allegations. There was no evidence that Dr Wong intentionally falsified T&E reports to charge his salary against the Federally funded grant, for work that he did not complete. Dr Wong made reasonably good progress on two of the four projects that he was assigned, and he received insufficient supervision and mentorship for the other two aims (and should probably have never been assigned to those projects in the first place). Rather, this is another case study, within this volume, to highlight the critical role—and potential shortfalls—of appropriate mentorship by senior researchers.

The RIO raised concerns that the center's leadership committee seemed to have let their dissatisfaction, regarding Dr Wong's work performance, remain unsettled for too long, and that he was not offered the mentorship he required over the past 15 months. It seemed likely that, at least in part, Dr Wong did not fully understand what the committee was asking of him, and he clearly did not possess all the skills required to meet every aim in his section of the grant. Moreover, it is quite possible that a significant portion of the confusion related to his work performance, or lack thereof, may have been due to a more fundamental language barrier.

## 6. RESOLUTION

By the time Dr Saltzman's complaint had been brought to the attention of the institutional RIO, the center's leadership committee had already decided to suspend all future funding to Dr Wong from the center grant. However, the Vice Chair of Surgery, after consultation with the RIO regarding the disposition of this case, agreed that the department should assume a portion of Dr Wong's salary to continue his employment (in a different lab) for another 12 months—thus allowing him a reasonable opportunity to seek new employment or additional training. The Vice Chair also felt that he had been performing in what he believed to be "good faith," and there is reasonable doubt regarding any breach of scientific integrity by him, and the RIO decided not to pursue this case any further.

## 7. COMMENTARY

This was an unfortunate case in which the individual accused of defrauding the US Government by inappropriately charging T&E to a publicly funded center grant, had no apparent understanding that he might have been doing anything that is ethically questionable. By the time this concern was raised with the RIO, the principal investigator and leadership committee for the center had all allowed their concerns of poor performance by Dr Wong to grow over a 15-month period, with little demonstrated effort to remediate the situation. In fact, it seemed quite plausible that Dr Wong presented as a convenient employee to fill a hole in a grant application, when the lead investigators were struggling to get their large and complex grant application by the NIH submission deadline for peer review. Once the center grant was awarded, and Dr Wong was plugged in to his four projects in the laboratory, he seemed to have been left to struggle without consistent and helpful mentorship by Dr Bartlett and Dr Saltzman.

Training as a scientist, at the graduate, postdoctoral, and even junior faculty levels is based on an apprenticeship model that traces its roots to medieval Europe approximately 800 years ago. It remains the case that the best, most effective method for training academics for research careers is the apprenticeship model, with untold hours of one-on-one supervision by, and discussion and debate with, an experienced senior scholar and mentor. This endeavor requires an intense amount of T&E to be expended by both parties—the mentor and trainee. An ideal mentoring relationship is one that is built on trust, mutual commitment to the educational process, and a sense of "calling" on the part of the mentor to pass on his or her skills, knowledge, and insight to the next generation of scientists who will, in turn, advance the leading edge of knowledge in a given field of study. Mentoring students is time-consuming and almost always leads to an expected decrease in work "efficiency" for the mentor. There are numerous publications, published guidelines, and even training courses to improve mentoring skills and to enhance the mentoring relationship (cf. Centering on Mentoring Presidential Task Force, American Psychological Association, 2006). In this case, it appeared to the RIO that this bond of trust, and training commitment, had not been attended to with enough diligence by Dr Wong's mentors; thus allowing poor communication and unreasonable expectations to sour what could have been an exciting start to Dr Wong's faculty career.

## QUESTIONS FOR DISCUSSION

*Parties*: Dr Tien Wong, defendant, young research assistant; Dr Matt Saltzman, complainant, Chief for Division of Pediatric Research; Dr Brian Forrester, assigned mentor/advisor for Dr Wong; Dr Jack Bartlett, PI for the center grant.

1. Does the RIO have any responsibility to address the mentorship lapses by Dr Saltzman and Dr Bartlett? Is there any accountability for Dr Saltzman regarding his choice of Dr Wong for the position?

2. Does the termination of Dr Wong (and the extension of his employment in another lab for 12 additional months) present any concerns or issues, given the findings of the RIO? If so, what are they?

3. This case (like many research integrity cases) presents issues of cultural misunderstanding. How can RIO address these issues proactively before research misconduct occurs?

4. The misconduct at issue here (possibly inflated invoices) appears to be outside the scope of the federal regulations' definition of research misconduct (prohibiting "making up data" and "changing or omitting data or results" among other things). How would you advise an institution to broaden or limit its definition of misconduct, within its local policy?

5. What role did the partial responsibility of other individuals (Dr Wong's supervisors) play in the decision of no research misconduct?

6. If you were the RIO charged with investigating this complaint, would you have any concerns that the principal investigator may have intentionally misled the NIH review panel with regard to the capabilities of the lab and institution to host the large center grant?

7. What structural, administrative, or oversight measures could be put in place, to ensure that such situations do not reoccur? How should quality of mentorship, within an academic medical center, be monitored?

8. How might Dr Wong have been better protected in the first years of his faculty appointment?

## REFERENCE

Centering on Mentoring Presidential Task Force, American Psychological Association, 2006. Introduction to Mentoring: A Guide for Mentors and Mentees. American Psychological Association, Washington, D.C. URL. https://apa.org/education/grad/intro-mentoring.pdf.

CHAPTER 7

# Submission of Fraudulent Data to a Peer-Review Journal: What Is the Role of the Lab Head/Mentor?

## 1. PRESENTING COMPLAINT

In May of 2011, the Chief of Developmental Pediatrics for our medical center was contacted by the Chair of the Publications Committee for the American Federation of Biological Societies (AFBS), to inform him of their decision to ban one of the postdoctoral fellows in his Division, Dr Suharto Alatas, from submitting any future manuscripts to the ten journals that are published by the member societies of this organization. This ban, placed on any research reports authored or coauthored by Dr Alatas, resulted from an internal investigation managed by the Publications Committee, which found evidence that Dr Alatas had fabricated data in a manuscript that had been sent for external peer review. The Division Chief had received a letter detailing this decision, which was then brought to the proper office within the medical center that is responsible for managing scientific integrity issues for the institution. The AFBS letter contained a request that Dr Alatas' home institution launch an internal investigation and to take any additional actions as deemed appropriate.

At the time that these events took place, Dr Alatas was employed by the medical center on a full-time basis, as a postdoctoral fellow in the Division of Developmental Pediatrics. He was working under the direct supervision of Dr Abdul Perkasa, a young Assistant Professor. Dr Alatas was Dr Perkasa's first postdoctoral fellowship student, and Dr Perkasa had agreed to be listed as a coauthor on the manuscript in question. This manuscript included photomicroscopic images of immunocytochemistry stained sections, and the Publications Committee of the AFBS had claimed that the one important image in the paper was entirely mislabeled, the Methods section described an analytic procedure that was not reflected in the photomicrographs, and the key image (see Figure 1(A)) had been intentionally fabricated by manipulating a comparison image using a photographic design program (i.e., Photoshop; Figure 1(B)).

*The Management of Scientific Integrity within Academic Medical Centers*
http://dx.doi.org/10.1016/B978-0-12-405198-0.00007-2

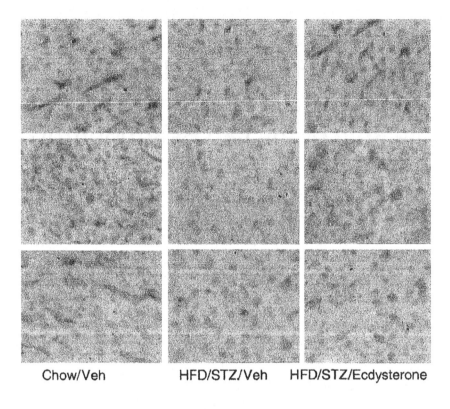

Chow/Veh          HFD/STZ/Veh    HFD/STZ/Ecdysterone

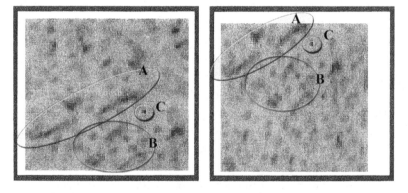

**Figure 1** (A) The nine-panel set of immunocytochemistry images, submitted with the original manuscript, to the editorial office of the American Federation of Biological Societies journal. These images show staining of the insulin receptor substrate 2 (IRS-2) in the livers of fetal rats for each experimental group. Each image was noted to have represented a different animal. (B) These are two panels from Figure 1(A) (top left and top right panels). The left panel is a stretched version of the right panel. Using an "overlay tool," the photographic image expert consultant was able to stretch the right panel to fit on top of, and to match precisely, the left panel. Sections A–C, in the above two panels, are identical point-to-point between these two images.

## 2. BACKGROUND AND HISTORY

Dr Suharto Alatas, an Indonesian national from Universitas Airlangga (Surabaya, East Java), came to America on a time-restricted visa to complete a postdoctoral fellowship. His primary research interests, centering on the embryology of the central nervous system, matched those of Dr Perkasa, thus making him an ideal mentor for Dr Alatas' fellowship. Because this was a topic that he had focused on during graduate school, he brought with him several folders of data and more than a dozen boxes of stained slides, so that he could finish a manuscript that he had started prior to departing Indonesia one year ago. Moreover, because Dr Perkasa was his new mentor, and he had asked his advice on multiple occasions about this manuscript, and because he also helped to correct his English-language grammatical and syntactical errors in the drafting the manuscript, he had invited Dr Perkasa to be a coauthor on this paper submission. It is important to note, though, that the manuscript described empirical research conducted in Indonesia, before moving to the United States to join Dr Perkasa's laboratory.

In January of 2011, Dr Alatas submitted his completed manuscript to the journal, and he received confirmation of its receipt. The manuscript was sent to three ad hoc peer reviewers, to solicit independent opinions about its suitability for publication. In early March, the Editor-in-Chief of the journal, Dr Richard Tracy, contacted the author to let him know that a portion of the Discussion section appeared to be missing, and that the results listed in an accompanying table for the Western blot analyses appeared to be unusual and not consistent with typical Western blot data (the listed standard errors of measurement were judged by the reviewer to be unusually small). As a result, Dr Tracy asked the author to provide the original blots for inspection. One week later, Dr Alatas responded to this request with an apology for any confusion, and he provided "the correct manuscript, including the complete Discussion section."

Several figures containing the Western blot and immunohistochemistry images were sent back to the three peer reviewers, and one of these individuals clearly inspected the images quite carefully. She noticed that on two immunohistochemistry images, representing differences between two experimental groups, the stained proteins seemed to have the exact same shape—which would be entirely unexpected (see Figure 1(B)). This reviewer called this oddity to the attention of the journal editor and, in turn, he hired a Photoshop image expert to analyze the two images in question, and to determine if it would be possible—in just a few steps—to take one image and to make it look like the other. The external consultant "reverse engineered" the second of the two images, and showed how it could be transformed into the first image in just a few steps using the basic tool set within the Photoshop program.

With the consultant's report in hand, Dr Tracy brought this case to the attention of the Publications Committee for the AFBS, for review and consideration. This committee, in turn, determined that there was sufficient evidence to believe that the data provided by Dr Alatas were fraudulent, in whole or in part, and that he should be sanctioned accordingly. In May of 2011, the Chair of the AFBS Publications Committee sent a letter to Dr Alatas, to inform him that the committee believed that the data contained in the manuscript(s) he had submitted were "misrepresented." The letter listed several specific concerns and findings. "First, it is not appropriate to cut molecular weight markers from one gel and paste them onto another, and yet represent the gels as independent. Second, your original manuscript indicated that the composite data on protein expression for the insulin receptor IRS-2 were based on analysis of independent immunohistochemical sections. However, when I reviewed your images of the various reportedly independent sections upon which your analyses are based, at least two of the six panels in your key figure appear to be unusually similar."

## 3. RELEVANT EVIDENCE

The manuscript in question contained the six-panel figure shown in Figure 1(A). These six panels show immunocytochemistry staining for the insulin receptor substrate 2 (IRS-2) in the livers of fetal rats for each experimental group. In the original manuscript, each of these images was noted to have represented a different animal. The Photoshop image expert, hired as a consultant by the AFBS journal editorial office, showed that the top left and top right panels in this figure are identical, that is, the top left panel is a stretched version of the top right panel (Figure 1(B)). This consultant applied an "overlay tool" to stretch the right panel to fit on top of, and to match precisely, the left panel.

In reviewing the supplemental material, the Inquiry Committee found still another example of using the same photomicrograph to represent immunohistochemical analysis of IRS-1 protein expression to represent two unrelated animals from different experimental groups. This additional instance of apparent photographic manipulation was also shown to

Dr Alatas, who claimed that he was not aware of that error and believed that it occurred the same way as the other one, i.e., by constructing figures using photomicrographs from a computer folder of images that were inadvertently mislabeled.

## 4. CASE MANAGEMENT AND ADJUDICATION

As noted above, a formal letter of complaint was sent by the AFBS to the Chief of the Division, within the medical center, responsible for Dr Perkasa's laboratory. This letter was delivered to the institution's Research Integrity Officer (RIO). The RIO took immediate steps to: (1) sequester all laboratory notebooks, slides, and other files owned by Dr Alatas; (2) place him on a temporary suspension from the laboratory, animal colonies, and other core facilities; and (3) form an Inquiry Committee consisting of four members. This committee was comprised of the Chair of the Department of Medicine, a senior-level scientist in Dr Alatas' area of specialization, a senior-level scientist in a completely different department and medical specialty, and a respected researcher from within the institution who is both a native Indonesian speaker and equally fluent in the English language.

Once all four members of the investigative committee agreed to serve the institution and accept this serious responsibility, the RIO charged the committee with the following request: "Approximately two weeks ago our Chief of Developmental Pediatrics received a letter from a journal editorial office, regarding a potential case of scientific fraud involving both a postdoctoral fellow in his Division, Dr Suharto Alatas, and his mentor—an assistant professor at this institution. This committee must review all of the manuscripts, source documents and data, and interview both Dr Alatas and Dr Perkasa, to determine whether—in your considered opinion—there is sufficient reason to suspect that an intentional act of scientific fraud has been committed."

Dr Perkasa was informed of this investigation and process, and although he was allowed to continue to work in the laboratory with his other students during this investigation, he was instructed not to submit any further grant applications to external agencies, nor any new journal manuscripts for external peer review, pending the outcome of this investigation.

By early June, 2011, the Inquiry Committee sent a letter to the RIO to inform him that the committee had drafted interview questions, and that separate meetings had been scheduled to interview Dr Alatas and Dr Perkasa. Selected portions of both interviews are provided below:

## 5. INTERVIEW WITH DR ALATAS

1. *When did you begin working for the hospital?*
I became affiliated with Dr Perkasa and the Division of Developmental Pediatrics when I answered a website posting in Indonesia advertising a position in Dr Perkasa's laboratory. I was able to interview with Dr Perkasa by video conference, and I was then offered a fellowship position and began working in Dr Perkasa's lab at the end of June, 2010.

2. *When was the work for your manuscript started, and when was it completed?*
The work described in that particular manuscript was done by myself in the laboratory of my graduate school mentor, in Indonesia, between 2007 and 2010.

3. *Where was the work for the manuscript done? Specifically, was work done here at this institution, in Indonesia, or in both places?*
All of the work was done in Indonesia, as part of my PhD thesis research. I completed additional experiments for two years after I had received my PhD degree, while still in Indonesia, but none of that work had been published when I had the opportunity to start my fellowship with Dr Perkasa.

4. *Was any of the work done on this manuscript started as a collaborative effort between you and Dr Perkasa, prior to your arrival in the United States?*
No. I approached Dr Perkasa about publishing this work in an American journal after I arrived in his lab. Dr Perkasa agreed to translate my manuscript into English, and he helped me with the Discussion section, but he did not perform any of the experiments and he did not analyze any of the data presented in the paper.

5. *Did you bring all of the data that was submitted for this manuscript with you from Indonesia or did you leave parts of it there?*
All of the data, as well as my notebooks, have been retrieved and are now in the possession of this committee.

6. *What were the roles of the coauthors regarding the data that were submitted as part of the manuscript?*
The middle two authors are my PhD mentors from Indonesia. Dr Perkasa, the last author, typed the manuscript by translating part of my thesis, he created a new figure, and he added his comments to the Discussion section.

7. *What was the initial response that you received from the journal editor, after you submitted the manuscript for peer review?*

They wanted the original data, to see why the standard errors were so small.

8. *Was there anyone else who contributed to the supplemental data that you submitted to the Journal?*

No. Dr Perkasa was on vacation when the request came. I contacted him and he told me that it was my paper and that I should send in whatever material I thought was appropriate.

9. *How many of the immunocytochemistry images in the submitted manuscript resulted from cutting and pasting of the molecular weight markers?*

The first two figures for the IRS-2 protein have pasted markers. The third figure is unaltered.

10. *Were there any other alterations made to these figures?*

No.

11. *You have a table showing no difference in molecular weights, and unusually low standard error of measurements, for two different proteins. How can this be?*

The actual molecular weight of IRS-2 is 291 bp. It was described erroneously in the paper and I did not report this mistake to the journal.

12. *Who was responsible for preparing and labeling the immunocytochemistry stained images for this study?*

Photographs of the gels of the PCR products were given to a statistical imaging center at my university. This center was responsible for generating image analyses data, statistical comparison, and sample images. They did the analyses and gave me the results.

13. *Were the protein levels measured by Western blots, as you stated in your submitted Methods section, or by immunohistochemistry—which was not mentioned in your paper?*

They were measured by digital analysis of immunohistochemistry sections. The detailed description of a Western blot technique in the Methods section and labeling of the data in the table, as being generated from Western blots, was provided in error. Dr Perkasa wrote the Methods section and legends for the tables and I did not see this error until the journal asked me to send them the original blots. Western blots were originally produced in Indonesia, and so I responded by sending the journal pictures of those blots. But, the data in the table still were based on image analyses of other immunohistochemistry sections.

14. *Why were the same images used to represent unrelated animals from different experimental groups, on two different figures within your manuscript?*
This occurred because the figures had been mislabeled. I submitted all of the photographs of immunohistochemistry sections to the image analysis department at my institution. They gave me printouts of the data from each photomicrograph and a table summarizing the means and SDs for each group. I recorded the mean and SD values for each experimental group in my notebook (this is demonstrated during the interview from one of the notebooks recently retrieved from his lab in Indonesia) but the original computer printouts showing the data from the image analysis center were destroyed. Long after the photos were analyzed, I constructed a figure that consisted of an immunohistochemistry photo from three animals in each experimental group stained for IRS-2 (see Figure 1(A), above). The figure was created from digital folders stored in a computer file. When it was pointed out to me by the journal editor that the same photo had been used to represent two animals from different experimental groups, I discovered that more than one copy of the photo existed in my computer file and it had been labeled incorrectly.

## 6. INTERVIEW WITH DR PERKASA

1. *When did you first meet or communicate with Dr Alatas?*
I met Dr Alatas when he responded to an advertisement for a postdoctoral position in my lab. I received about 20–30 resumes in response to an advertisement, and I felt that his background was best suited for the work I needed done. I spoke personally with three references of his, and they all spoke very highly of Dr Alatas, and so I decided to offer him the position.
2. *What was the nature of your working relationship with Dr Alatas?*
I hired Dr Alatas as a postdoctoral fellow in my lab. He was 2 years out from his PhD, but this was the first postdoctoral position that he had taken.
3. *Were any of the data that he submitted to the AFBS journal your own data?*
No.
4. *What was your role as the senior author on the paper?*
When I read his thesis, I completed additional background reading and I began to formulate a hypothesis that might explain the results that Dr Alatas had obtained. I translated a fraction of his doctoral thesis that was written in Indonesian, into an English manuscript, and I added some of

my own ideas about the results that had been obtained. These ideas were expressed primarily in the Discussion section and with a new figure that I had created. I asked Dr Alatas to proofread the English manuscript that I had written, based on his experiments, to make certain that I had accurately described his work. He told me that he had checked the manuscript, as I had requested, before it was submitted to the journal. I felt that it would me much easier for me to write the manuscript in English than it would be for him, but I felt that his English was more than adequate to proofread the paper.

5. *Did you see the reply that was written by the journal's editor to Dr Alatas, soon after his manuscript had been received?*

No. I was on vacation when the request for a reply was received by Dr Alatas.

6. *Did you help Dr Alatas to prepare a written response or the supplemental materials for the journal?*

No. I was on vacation.

7. *Have you seen the supplemental data that was submitted by Dr Alatas and, if so, what is your interpretation of that data?*

No, I have not seen his response to the initial letter.

8. *Were you aware that Dr Alatas had altered the figures he submitted to the journal by pasting incorrect molecular markers on the gels?*

No.

9. *Were you aware there were discrepancies in the migration lengths of the PCR products?*

No.

10. *Did you know whether Dr Alatas used Western blots or immunohistochemistry to quantitate protein levels?*

No. I only know two methods for protein quantification: Western blots and ELISA immunohistochemistry. Since Dr Alatas' thesis contained a Methods section on Western blotting, I just assumed that he used that method for quantification.

11. *Were you aware the photomicrographs from the same animal were used to represent animals in different groups?*

No.

12. *Did you advise Dr Alatas on how to reply to the AFBS's ethical concerns?*

Yes. I told him to reply immediately and explain as best he could what had happened.

13. *Did you seek advice from anyone else on how to respond to the concerns of the journal and its editorial office?*

No. I had not experienced allegations of misconduct in research previously and was not aware of anything that needed to be done other than responding to the concerns raised by the journal.

## 7. SYNTHESIS AND FORMULATION

After reviewing all of the English materials available to the committee, and after conducting interviews with both Dr Alatas and Dr Perkasa, the Inquiry Committee determined that the events related to the concerns raised by the AFBS could be summarized as follows.

1. Although in the manuscript Dr Alatas listed both his prior affiliation at the Universitas Airlangga, as well as his current home institution, all of the data submitted in the manuscript were derived from experiments performed while he was in Indonesia and analyzed by himself or by personnel in departments at his former institution that specialized in image analyses. Dr Perkasa did translate the manuscript from Dr Alatas' PhD thesis, and he also elaborated on a potential model to explain the experimental results.

2. All of the concerns raised by the AFBS are appropriate given the original manuscript, revised manuscript, and the supplementary materials that were provided to them by Dr Alatas.

3. There were four major discrepancies between the final written work that had been submitted for peer review, versus the information and raw data collected by Dr Alatas in the course of his research:

   a. The data presented in the primary data table, represented in the paper as the results of measuring specific protein levels by Western blot technique, were actually derived from computer analyses of photomicrographs of immunohistochemistry sections. This technique is generally considered to be inferior to the preferred Western blot technique. The writing of a paragraph entitled "Western blot analysis" in the Methods section, the lack of any description of immunohistochemistry methods, and the labeling of data in the table as being derived from the former approach is a clear misrepresentation of data.

   b. The lack of variation, and low values, for the standard errors of measurement reported in the data table, noticed by the reviewers, was due primarily to a statistical error by Dr Alatas (he had divided the standard deviations by the wrong denominator in the equation). This occurred from pooling mRNA from nine animals into three PCR amplifications. This error would not have affected the overall findings of the study, but could have helped to make the between-groups differences seem more statistically significant.

c. Molecular markers, in different figures, were digitally cut and pasted onto four of the 12 protein-stained panels submitted to the journal as supplementary data in support of the manuscript. These two manipulations resulted in a misrepresentation of what the reviewers were looking at, in that it prevented them from accurately determining whether the PCR products generated from each experiment were likely to be those claimed by the authors.

d. Immunohistochemical stains, for two different proteins, were used from single animals in order to represent differences between experimental groups. The same appeared to happen in two separate instances within the same manuscript. In addition to the first error identified by a peer reviewer for the journal, the Inquiry Committee found a second such misrepresentation of visual data. The committee was particularly concerned that, in both instances, these errors resulted in the "treatment group" looking more like the healthy "control group" and thereby supported the authors' major claim regarding the timing of IRB-2 protein signaling on embryological growth of the cerebral cortex. Moreover, there was no way for the committee to ascertain whether these errors in visual representation of data for the two groups carried over to the accompanying data table, because the images were labeled for use in the figures by Dr Alatas long after they had been digitally quantified by a statistical center at his prior institution in Indonesia. However, the committee was of the opinion that it is difficult to assume that the analyses of the two different proteins would be correct for the tabular presentation of data, but incorrect for the figures, especially as both errors in the figures resulted in visual support for the author's primary hypothesis.

4. Dr Perkasa made several important factual mistakes in his translation of Dr Alatas' work. It was apparent from interviewing both authors that Dr Perkasa considered the manuscript to be work done entirely by Dr Alatas before he came to work in the laboratory, and as such, subject to little review and supervision by him. It is also clear that Dr Perkasa relied heavily on Dr Alatas' final proofreading of the manuscript that he had translated from Indonesian into English, to ensure that what he had written accurately described how the experiments were performed and how the results were analyzed.

5. In proofreading the manuscript, Dr Alatas either failed to notice—or chose to overlook—a critically important error that occurred during the translation of his thesis into English. Namely, the protein levels listed in the main data table were not derived by Western blot (which is a more accurate and preferred method, and would certainly be viewed as such by journal reviewers), but rather by digital analyses of immunohistochemistry-labeled sections. This error occurred in two different parts of the manuscript (an entire portion of the Methods section, and the legend that accompanied the main data table). Although the Inquiry Committee could understand how Dr Alatas might have overlooked the legend for the table, it was almost inconceivable to the committee that he could miss an entire paragraph in the Methods section describing the procedure for Western blots that were never used to generate data for this manuscript. Furthermore, no description of the immunohistochemistry technique(s) was provided in the Methods section.

6. Neither Dr Alatas, nor his supervisor Dr Perkasa, was familiar with policies that outline responses to an allegation of research misconduct. Neither investigator informed their superiors, employers, or other members of the Department of Medicine.

## 8. RESOLUTION

The RIO for the medical system took appropriate steps, with the faculty member involved (Dr Perkasa) and with his Division Director, to ensure that none of the data in the manuscript under investigation would be included in any future articles or grant applications of any type. In addition, a second submission of this same work, by Dr Alatas to an unrelated journal, was withdrawn from further consideration. No portions of this manuscript have been submitted elsewhere for review and publication.

The RIO prepared a final report to the AFBS, to inform the organization that an internal review of the matter had been concluded, and the findings were in support of their prior concerns and decision to bar Dr Alatas from submitting new manuscripts to the group of scientific journals that they manage.

Based on the review and findings of the Inquiry Committee, the RIO determined that there was sufficient concern that Dr Alatas intended to publish fraudulent experimental results, although there remains the possibility that the series of multiple mistakes (all favoring support for the author's primary hypothesis) could have resulted from a pattern of poor laboratory practice and poor oversight by his mentor. The RIO had recommended to the institution's Deciding Official that Dr Alatas' employment be terminated, due to inappropriate professional conduct and judgment, but as this case was coming to a close, Dr Alatas made an independent decision to resign his postdoctoral fellowship and to return to his home country.

As this case was coming to closure, the RIO met with the "Deciding Official" for the institution (who also served in the role of Chief Medical Officer), to appraise her of the case, its handling, and likely final disposition. The Deciding Official strongly believed that Dr Perkasa's employment as a faculty member in the institution be terminated, on the grounds that he failed to adequately oversee the professional conduct of a postdoctoral fellow in his laboratory, and that he allowed himself to be listed as a senior author on a manuscript for which he had not inspected the quality and veracity of the experimental results. The RIO, in turn, pushed back on this request and made the point that this was the first time that Dr Perkasa had been supervising a postdoctoral fellow, that he was new to that role and had much to learn about mentorship, but that there was no direct evidence that Dr Perkasa was himself complicit in any intent to publish fraudulent data. Dr Perkasa otherwise had a completely unblemished record and was judged by all accounts to be at the start of a successful and productive scientific career.

On further discussion, the Deciding Official agreed to allow Dr Perkasa's employment to continue, on the condition that he meet regularly with the RIO for a series of "coaching sessions" to better understand the principles and practices of effective mentorship for graduate students and postdoctoral fellows. These coaching sessions occurred regularly for one year, with Dr Perkasa noting on more than one occasion that he had "learned a hard lesson" and would be far more careful with respect to all intellectual work coming from members of his lab group, not to mention being more careful about what publications he would agree to have himself listed on as an author. Since these events took place, Dr Perkasa has successfully competed for several large grants, he has received an early career research award in his field of study, as well as a teaching award within the medical school. He now manages a larger, active laboratory group with several graduate students and postdoctoral fellows.

## 9. COMMENTARY

It remains unclear as to whether Dr Alatas had specific, premeditated intentions to commit scientific fraud, or whether the multitude of errors in his work—all erring in the direction of supporting his primary hypothesis and chances for favorable reviews and publication—were the result of unusually sloppy laboratory and data management conduct. In either case, he was deemed to be unsuitable for continuation of training as a postdoctoral fellow. A few days prior to any action being taken, with respect to his employment, Dr Alatas chose to resign voluntarily and to return to his home country. His ban on future publications in the ten journals managed by the AFBS remains in effect by that professional society. Although we will never know his personal motivations to consider misrepresenting data—if he, in fact, intended consciously to do so—it should be noted that there are likely a variety of societal pressures and sources of secondary gain, with respect to guaranteeing successful publication(s) that Dr Alatas was aware of. As an example, it is exceptionally difficult for Indonesian PhD scientists to obtain permanent teaching/research positions in their home country, and obtaining a job in his chosen field would be essentially impossible without having first completed a postdoctoral fellowship (preferably in North America) and publishing articles in English-language peer-reviewed journals. Hence, it is quite likely that Dr Alatas was under intense pressure to publish several articles while studying in the United States, in order to have any measure of success in finding suitable employment on his return to Indonesia. At any rate, Dr Alatas chose to voluntarily resign his position before the institution terminated his employment.

The more difficult issue that the RIO faced in this case did not involve the postdoctoral fellow, but rather, his supervising faculty member. There was strong pressure from the senior administration at the medical center to terminate Dr Perkasa's employment so that the institution might distance itself from the appearance of tolerating scientific fraud. It is sometimes the case that research institutions take strong preemptive steps, including the termination of employment for a variety of individuals, in order to rapidly and cleanly remove any possibility of associating the larger institution with claims of scientific misconduct (which could easily wind up in the public press). However, in this case, there was really no objective evidence suggesting that Dr Perkasa knowingly engaged in misconduct. Rather, he made several errors in judgment that, although relatively common in young faculty who are inexperienced mentors, are made very frequently by junior and senior academics alike. First, Dr Perkasa allowed himself to be listed as a senior coauthor on a paper detailing experiments that he was not personally involved with, did not contribute to the design of, and for which he did not take time to understand the mechanics of at a deep level. For instance, Dr Perkasa was unaware that the resulting data from these experiments were analyzed using two completely different protein quantification techniques (with the data resulting from a less-desirable technique being represented in the paper as resulting from a technique that would be preferred by peer reviewers). In addition, Dr Perkasa clearly failed to ensure that Dr Alatas had proofread the translated manuscript for accuracy, despite the fact that he was a nonnative English speaker and probably required assistance in this regard. At the very least, Dr Perkasa should have met together with his fellow, for as long as required, to complete line-by-line editing and proofreading of the translated manuscript. Unfortunately, this level of oversight by Dr Perkasa never took place, probably because the experiments detailed in the paper were not his to begin with, and he did not see the need to carefully supervise work that did not originate in his own laboratory.

The question should be asked, then, whether Dr Perkasa was justified at all in agreeing to have his name listed on the paper as a coauthor. There is a large debate in science, and among those who consider the research ethics, as to what types and levels of individual contributions are sufficient for those who are listed on any given article as coauthors. There is no doubt that many senior investigators have had their careers ruined (or nearly so) after insisting that their names appear on publications for which they had no direct oversight or responsibility. As an example, in the early 2000s the then powerful chairman of Obstetrics and Gynecology at Columbia University School of Medicine in New York City, Dr Richard A. Lobo, resigned following an embarrassing scandal involving fraudulent research published—with his name as the senior author—in a top journal for which he also served on the editorial board (Cho et al., 2001; see discussion by Snyder and Mayes, 2009). Another example might be the legal case brought against a junior faculty member at MIT, by the Office of Research Integrity of the NIH in the late 1980s, and that involved published research coauthored by the biologist and Nobel Laureate, David Baltimore. That high-profile scandal nearly derailed Baltimore's career, forcing him to resign as the President of Rockefeller University in the early 1990s—although he was later acquitted of all charges. Unfortunately, there are all too many similar examples to cite, and these cases garner media attention with great regularity.

Of course, in the case of Dr Perkasa, the circumstances were quite different. He did not insist that his name be added to the manuscript simply because he was the head of the laboratory that employed Dr Alatas. Nonetheless, Dr Perkasa did not contribute substantially to the paper, but rather assisted his trainee without taking the time to fully immerse himself in the details of the project, in order to assist Dr Alatas in publishing his older work that he brought with him from his PhD program. Dr Perkasa did not provide direct oversight of the image analyses, and he did not ensure that these methods and the handling of the images were all reported faithfully in the paper and its accompanying figures. This raises the obvious question, "What are the boundaries of coauthorship?" Do all coauthors need to hand-check the veracity of all individual data points included in an empirical report?

Thinking back to an example raised above, a few years after the scandal that nearly ended Dr David Baltimore's career, he accepted the post as President of the California Institute of Technology (CalTech). However, in 2005 Baltimore stepped down from this position amid new allegations that he may have been involved in yet another data fraud case involving a former trainee in his laboratory. Two years later, a scientific integrity committee at CalTech determined that his former postdoctoral fellow committed scientific fraud but acted alone—and, again, Dr Baltimore was cleared of all suspected misconduct. As this sort of event happened twice in his career, was Dr Baltimore just unlucky? Throughout the majority of his career, Dr Baltimore has overseen massive research programs and very busy laboratories. He is personally responsible for the training of dozens of talented and ambitious graduate and postdoctoral students. How can anyone directly oversee the accuracy of all data being collected, analyzed, and published by so many people at one time? And yet, in the end, all of that work is completed because of his mentorship and financial support (via grants he receives), and his name as the senior author on so many publications signals to the scientific world that the paper is a product of his laboratory. Any senior scientist, typically the head of a laboratory, faces this problem. Although David Baltimore is an extreme example, Dr Perkasa had a single postdoctoral fellow who was eager to complete and publish work that he brought with him, and Dr Perkasa did not directly check each stained slide, each quantified image analysis, or trace the origin of every stain in the figures chosen for publication. To accomplish this level of oversight would have taken several days, if not weeks, of valuable time that Dr Perkasa would prefer to devote to current work in his laboratory, to his own writing, and/or to the preparation of new grant applications. This decision to provide less-than-full supervision, in the preparation and publication of this paper that bore his name as a coauthor, was a mistake that could have ended Dr Perkasa's career at this institution.

## QUESTIONS FOR DISCUSSION

Parties: Dr Suharto Alatas, defendant; Dr Abdul Perkasa, Assistant Professor, Dr Alatas supervisor, coauthor on manuscript; Dr Richard Tracy, Editor-in-Chief, reported case to AFBS.

1. How do you think the accusatory letter from the journal affected the Inquiry Committee? Did it predispose the committee to find that Dr Alatas committed misconduct? Could or should steps have been taken to lessen any prejudice?

2. Neither Dr Alatas nor Dr Perkasa was aware of the institutional policies regarding research misconduct. Should more rigorously applied procedures be developed to ensure that researchers are aware of these policies? What kind of procedures, and how should they be communicated?

3. Was there any investigation after the institutional inquiry? Or was this step rendered unnecessary by Dr Alatas' resignation?

4. Neither Dr Alatas nor Dr Perkasa asked to be represented by a lawyer during the institutional proceedings? If they had, would that have changed anything?

5. Does the institutional pressure to separate itself from any perceived misconduct create a conflict of interest? If so, does this conflict call into question the Federal government's decision to place the primary responsibility for addressing research misconduct at the institutional level? If institutions perceive the potential for such conflicts, how should they deal with them?

6. In light of the fact that institutions are free to define research misconduct more broadly than Federal regulations, do you think institutions should consider expanding their policies to cover insufficient oversight of subordinates?

7. Does using old work from another lab violate any ethics/accepted practices, similar to the debated practice of publishing multiple articles on one set of research data? Could it be research misconduct? And, at what point does a claim of coauthorship become illegitimate based on lack of work/contribution? Is exaggerating involvement in authorship itself a form of research misconduct?

8. Dr Perkasa allowed Dr Alatas to publish data collected at his prior institution, while he was working full-time in the Perkasa lab. To what extent should Dr Perkasa have overseen the quality and veracity of Dr Alatas' completion of this prior project?

9. Dr Perkasa did allow himself to be listed as a coauthor on the submitted manuscript. While it is true that he worked with Dr Alatas to improve the quality of the writing, and to re-think and improve on the conceptual arguments being made in the article, he did not review all the data and analyses himself. To what extent was Dr Perkasa, or any coauthor, personally responsible for the veracity of very complex information being presented in a new manuscript submission? Should Dr Perkasa have been more careful, or was it simply too much to expect of him to personally check each figure, image, and data point?

10. The hospital executive leadership strongly urged the RIO and Chief Research Officer to terminate Dr Perkasa's employment. If the Chief Research Officer had agreed with this course of action, would this be appropriate and defensible?

## REFERENCES

Cho, K.Y., Wirth, D.P., Lobo, R.A., 2001. Does prayer influence the success of in vitro fertilization – embryo transfer? J Reprod Med 46, 781–787.

Snyder, P.J., Mayes, L.C., 2009. Chapter 1. Introduction: the ethics of scientific disclosure. In: Snyder, P.J., Mayes, L.C., Spencer, D.D. (Eds.), Science and the Media: Delgado's Brave Bulls and the Ethics of Scientific Disclosure. Elsevier, Amsterdam, pp. 1–9.

## CHAPTER 8

# Plagiarism versus Data Shared between Junior and Senior Investigators

## 1. PRESENTING COMPLAINT

Doctors Marcus Elliot and Jean Smith and their colleague Dr Lester submitted a manuscript describing work they had accomplished when working in a lab together at their previous institution, IU. They did not submit the manuscript until joining their current institution. The list of authors included their names as well as a colleague at their current institution, Dr Tatum, who provided editorial assistance. Their former mentor and lab director, Dr Norris, was not included as an author and on seeing the published manuscript filed a complaint with their current institution alleging plagiarism.

## 2. BACKGROUND AND HISTORY

Dr Elliot and Dr Smith were close colleagues who had worked together in their previous laboratory with their senior research advisor, Dr Arnold. Dr Smith was especially technically skilled in microsurgeries in the mouse and Dr Elliot was a talented gastroenterologist interested in pancreatic function and the regulation of inflammatory responses in the pancreas. Dr Smith had obtained her doctorate earlier and had moved to Dr Norris lab. She encouraged Dr Elliot to follow her there and to complete his dissertation research with Dr Norris. Dr Smith and Dr Elliot were excited to obtain employment in Dr Norris' laboratory for he was an internationally recognized scientist focused on the immunology of the gut and the impact of stress on gastric inflammatory response. Dr Elliot had worked on pancreatic inflammation and the biology of pancreatitis in Dr Arnold's lab as part of a masters' thesis and brought these interests to Dr Norris' research group. Dr Smith and Dr Elliot had already published several manuscripts. Dr Smith was first

*The Management of Scientific Integrity within Academic Medical Centers*
http://dx.doi.org/10.1016/B978-0-12-405198-0.00008-4

author on four manuscripts from her dissertation and had a very collegial and ongoing collaboration with her dissertation advisor, Dr Arnold. She was also focused on the biology of the pancreas and had helped Dr Elliot when he was working in Dr Arnold's lab for his masters' thesis. Dr Elliot was a coauthor on one of Dr Smith's papers and also on two other papers in Dr Arnold's lab. Dr Smith and Dr Elliot continued to be in regular touch with Dr Arnold who had helped them find their jobs with Dr Norris and indeed Dr Elliot had decided to continue with Dr Norris work he had started with Dr Arnold.

Dr Norris was a senior investigator with a very productive laboratory. On average, he had two to three doctoral students and a similar number of postdoctoral fellows. His students and postdoctoral fellows worked well together and regularly found either faculty jobs in graduate and medical schools or in biotechnology firms. The lab published at least ten papers a year consistently in high-impact journals. Even though his laboratory was large with many lines of research ongoing, some of which while complementary were not directly related to Dr Norris' own research focus, his general approach to publications from the lab was to insist on being the senior author. This "rule" was understood by most members of the lab to apply even after they left the lab if they were completing manuscripts from data they collected while they were in Dr Norris' laboratory. As Dr Norris had become more senior and more distant from the work of many of his younger colleagues even though he supported their work through making available his laboratory facilities, there were subtle but growing tensions in the laboratory about Dr Norris' claim as senior author on all manuscripts he perceived as emanating from his research group.

As Dr Elliot successfully completed his dissertation, both he and Dr Smith were invited to join Dr Tatum's laboratory at another institution, WU. Dr Tatum was especially eager to have Dr Smith join his research team because of her considerable laboratory skills and generously helped Dr Elliot find a postdoctoral position in a related gastroenterology laboratory. Shortly after their arrival, Dr Smith and Dr Elliot presented Dr Tatum with a manuscript describing two experiments manipulating pancreatic inflammation. They included Dr Arnold as an author and told Dr Tatum that the work in the manuscript was initiated in Dr Arnold's lab several years earlier and completed in Dr Norris' lab as a part of Dr Elliot's dissertation. They had not included Dr Norris as an author and implied that he was not involved in the work and hence would not expect authorship. They asked Dr Tatum to help them edit the manuscript and offered him a place among the authors for his assistance. Only a few weeks after arriving at their new institution, Doctors Elliot, Smith, Arnold, and Tatum submitted the manuscript to a high-impact journal for review with Dr Elliott as the first and corresponding author.

Two months after submission, Dr Tatum received a letter from an administrative official at IU informing him that Dr Norris had submitted an allegation of plagiarism toward him for his involvement as an author on work that emanated from Dr Norris' laboratory. Dr Tatum immediately brought the letter to the Research Integrity Officer (RIO). Because IU was alleging Dr Tatum as a faculty of WU had committed plagiarism on a manuscript submitted with two new junior faculty members of WU, Doctors Smith and Elliot, along with Dr Arnold from another institution, the RIO convened an inquiry committee to review the matter. Dr Elliott was advised to immediately withdraw the manuscript from the journal's review process.

## 3. RELEVANT EVIDENCE

The inquiry committee requested that Dr Tatum supply all versions of the manuscript including the initial draft provided to him by Dr Smith and Dr Elliot and the versions showing his edits. They also asked Dr Smith and Dr Elliot to provide copies of any drafts they had completed prior to leaving IU and also asked Dr Arnold to provide all edited versions of the manuscript showing his input as well. The committee also established that the study in question as well as Dr Smith and Dr Elliot's salary support was accomplished without NIH funding. There was a question of whether or not NIH funds were used to purchase some animals and laboratory supplies used in Dr Elliot's research.

## 4. CASE MANAGEMENT AND ADJUDICATION

The inquiry committee interviewed Doctors Smith, Elliott, and Tatum. The committee also interviewed Dr Norris and Dr Arnold by phone. While there was some inconsistency in their statements, Dr Elliot and Dr Smith described their work on the studies represented in the manuscript as happening predominantly in the Norris lab. Dr Elliot implied that he had designed the study with Dr Arnold but it was not clear from his statement whether or not he had actually started experiments in Dr Arnold's lab. Dr Smith was clearer in her statements that all work was accomplished in the Norris lab. Dr Elliot and Dr Smith both contended that save for providing them laboratory space, animals, and reagents to accomplish the experiments, Dr Norris knew little to nothing about their work. They were unclear about whether or not Dr Norris had seen a draft of their manuscript before they left for WU but they were very clear that they did not agree with Dr Norris' insistence on being an author on any work emanating from his laboratory. Dr Arnold also spoke about his concern regarding how much Dr Norris was involved in Dr Elliot's research. Dr Arnold stated that he had reviewed Dr Elliot's data, the interpretation of the findings, and assisted in preparation of the drafts prior to submission for publication.

In contrast, Dr Norris stated that he had been actively involved in the design of the study described in the manuscript and had reviewed with Dr Elliot the findings for his dissertation that ultimately were included in the manuscript. Dr Norris contended that he was included as senior author in a very early draft of the manuscript and presented a first draft with his name listed. He provided a copy of that first draft with his marginal notes that he stated he had provided to Dr Elliot and Dr Smith. Dr Norris stated he did not know what happened to the manuscript after Dr Smith and Dr Elliot moved to WU. He assumed they would send him a near-final draft for his approval prior to submission for review. He was shocked when he was contacted by a journal editor who had received the manuscript stating in an acknowledgment that portions of the work were done at IU in the Norris lab but without Dr Norris being named as an author. The journal editor had been a postdoctoral fellow in the Norris lab a decade earlier and was hence familiar with the lab's work. Dr Norris was especially angry that Dr Tatum was named as an author for he assumed since they were working in similar areas, Dr Tatum was inappropriately taking credit for work done in Dr Norris', not Dr Tatum's, laboratory.

Dr Tatum stated that he was aware that Dr Smith and Dr Elliot had worked in Dr Norris' laboratory and Dr Elliot had accomplished his dissertation in the Norris lab. He understood from both Dr Smith and Dr Elliot that they had interacted only with a senior postdoctoral fellow in the Norris lab and had infrequent contact with Dr Norris. He understood that they had written the manuscript after leaving the Norris lab. His input had been primarily editing for grammar and clarity. He had not reviewed the data or analyses.

While the statements from the persons involved were conflicting, the committee felt there was sufficient reason to conclude that the research reported in the manuscript in question was accomplished almost entirely in Dr Norris' laboratory. Further, while Dr Elliot spoke about his research with

Dr Arnold, the committee concluded it was likely that none of the work had occurred prior to Dr Smith and Dr Elliot joining the Norris lab. The committee also felt that with an earlier draft of the manuscript from Dr Norris, it seemed clear there had been discussions about the work with Dr Norris prior to Dr Elliot and Dr Smith leaving his lab and that his name had been included as an author. Hence, the inquiry committee felt there were reasonable grounds to be concerned that the submission of the manuscript for publication without consulting or including Dr Norris as an author potentially constituted plagiarism. The committee felt Dr Tatum was not involved in plagiarism though they were critical of his decision to accept authorship for simply editing the manuscript without reviewing the primary data and analyses. The inquiry committee referred the matter to a second committee for additional investigation and included Doctors Smith, Elliot, and Arnold, but not Dr Tatum, as respondents. The committee advised the dean to convey to Dr Norris that Dr Tatum was no longer a respondent but that WU would further investigate Doctors Elliot's, Smith's, and Arnold's actions.

In the investigation phase, the committee gave Doctors Smith, Elliot, and Arnold the opportunity to review the findings and report of the Inquiry Committee and to present corrections and/or additional information. While none offered changes to the committee's report, Dr Smith and Dr Elliot strongly emphasized again how little support they had received from Dr Norris and how insignificant was his involvement in their work and the manuscript. In their subsequent interviews with the investigation committee, they acknowledged that Dr Norris had provided some comments on a very early version of their manuscript and that he was included as an author per his requirement. However, he had not reviewed any data or analyses and had not created or reviewed any figures. They also emphasized that the manuscript they submitted once they reached WU was substantially different from the earlier draft commented on marginally by Dr Norris. They were emphatic in their conviction that Dr Norris' "rule" about his inclusion as senior author on all work from his lab was both unfair and unethical. In contrast, Dr Arnold had reviewed all their data, analyses, and assisted in creating some figures from the data. He had been substantially helpful in their interpretation and discussion of findings and had made significant comments and edits on the version of the manuscript that was finally submitted for review.

Dr Norris spoke again to the investigation committee and was unable to provide any additional evidence that he was more involved in the study than the first draft with his few marginal comments. While he contended that he was well aware of Dr Elliot's work and he was the named advisor for

Dr Elliot's dissertation, there was no clear evidence to substantiate how involved he had actually been in Dr Elliot's work in the lab. Most of the day-to-day support for Dr Elliot's work seemed to be with the senior post-doctoral fellow who also served as the lab manager. It also became clear from Dr Norris' statements that he had not wanted Dr Smith and Dr Elliot to leave the lab, and that he was opposed to their submitting the work in the first draft for publication. He felt there was a need for more experiments and was not willing to sign off on a submission of the manuscript. Hence, neither Dr Smith nor Dr Elliot did any further work on the manuscript until they left Dr Norris' lab and arrived at WU nearly two years after their work on the research reported in the manuscript.

Because of the question of whether or not NIH funding was involved in purchasing some supplies and animals that were used in Dr Elliot's dissertation research, the NIH Office of Research Integrity (ORI) was informed of the investigation. On review of the investigation committee's findings, ORI declined further oversight of the matter because it was, in their view, an authorship dispute among collaborators.

## 5. SYNTHESIS AND FORMULATION

In contrast to the Inquiry Committee's formulation, the investigation committee concluded that Doctors Smith, Elliot, and Arnold had not plagiarized work from Dr Norris' lab and that there was no evidence that Dr Norris had contributed sufficiently and meaningfully to warrant authorship. The manuscript submitted was substantially different from the first draft. It also appeared that the work was entirely accomplished by Dr Smith and Dr Elliot from the original experiment design through data collection, analysis, interpretation, and presentation in the manuscript. While Dr Smith and Dr Elliot used the resources of Dr Norris' lab and were his employees/students, the committee felt he had been only distantly involved and his comments on the first draft were minimal and not central to the research findings and interpretation. Dr Arnold's involvement seemed more in line with responsible authorship. There was evidence that he had substantially commented on sequential drafts of the manuscript, had suggested additional analyses, and created at least one of figures.

While it is clearly difficult to accurately reconstruct the nature of individuals' relative contribution to any manuscript (especially when there are multiple authors), responsible authorship requires involvement at multiple levels of manuscript preparation. These include the design and implementation of the experiment(s) or study, processing and analyzing data, interpreting the findings, and preparing the manuscript and discussion of findings. Each laboratory has its own individual culture and allocation of responsibilities and efforts. Dr Norris asserted his ownership of all work by those in his laboratory and organized the lab hierarchically. Understandably the more experienced researchers in the lab, who were well on their way to independent lines of research, were less willing to automatically include Dr Norris as senior author, especially when he was minimally involved in their work. Further, in this case, Dr Norris refused to allow Dr Smith and Dr Elliot to submit the manuscript and delayed their work by requiring additional experiments as a condition for even considering completing and submitting the manuscript. Hence, Dr Elliot and Dr Smith put the work aside and waited until they had moved to another institution when they could revisit the data and manuscript without Dr Norris' oversight and control.

## 6. RESOLUTION

The investigation committee recommended that Dr Elliot and Dr Smith submit their manuscript for review for publication. Their report was provided to Doctors Elliot, Smith, Arnold, and Norris. Neither the complainant nor respondents provided any additional comments. As stated above, ORI declined to review the matter further as it was primarily a dispute among collaborators about authorship.

## 7. COMMENTARY

A core issue of this case is how authorship is determined and the responsible conduct of authors. It is not uncommon in some well-established laboratories for the senior and founder of the research group to imply, if not explicitly require, authorship on all work emerging from the lab. This practice is, in part, a measure of the shift from single authorship to multiauthored teams (Greene, 2007) reflecting complex collaborations and the increasing specialization of scientists and research teams (Price, 1986). It may also reflect the impact of how scientists are rewarded and evaluated in most contemporary academic systems such that multiple authors may claim the work in the manuscript as reflecting of their individual scientific productivity.

With multiauthored papers, several variations in authorship are more common. For example, the so-called honorary authorship is granted to an individual who played an insignificant role in the work reported in the manuscript (Greenland and Fontanarosa, 2012). This practice may be especially widespread even though the National Academy of Science (1995) cautions how honorary authorship dilutes the credit due to those who actually accomplished the work and makes it especially difficult to understand who is truly responsible for reported research. Nonetheless, senior colleagues may claim the role of supervisor or advisor without providing more concrete or measurable input to the work. Guidelines such as those suggested by Nature in which every author must outline their specific contributions may somewhat reduce this practice but by no means prevent it. Indeed, a survey of NIH-funded investigators found that 10% claimed to have inappropriately assigned or accepted authorship (Martinson et al., 2005).

Whether or not Dr Norris' expectation of senior authorship on all work emanating from his laboratory violates standards of responsible conduct in research is one matter. But the other core issue raised by this case is the question of plagiarism. While the definition of plagiarism may vary in subtle ways across institutions, it is generally viewed as the wrongful appropriation of another's "language, thoughts, ideas, or expressions" and the representation of them as one's own. Does exclusion of an individual who made significant contributions, however defined, to the work represented in the published report constitute an appropriation of another's ideas with false attribution to one or more other individuals as representing their work only? In the present case, there is a dispute about whether or not Dr Norris made significant contributions to warrant authorship, that is, how much he actually contributed to the design, data collection and interpretation, and manuscript preparation, the necessary contributions for authorship. Still, the question raised by this case and debated by both the inquiry and investigation committees (with differing opinions rendered by each) regards whether it is an act of plagiarism to exclude a contributing individual from authorship.

The present case illustrates a third issue around communication. Dr Norris had expressed his opinion on a much earlier draft of the manuscript that more data were required and he would not sign off on submission of a report with the findings as these stood. Dr Smith and Dr Elliot elected not to engage in further discussion with Dr Norris and instead to submit a much revised manuscript but without any new data as recommended by Dr Norris in his one set of comments on the very early draft. While perhaps understandable, they had included him as an author on the very early draft but did not have further discussions with him about his opinions or their decision to remove him from the list of authors on subsequent submissions. Setting aside the tensions in the lab around Dr Norris' approach to authorship or the question of how substantive was his contribution, is it appropriate to remove an author without discussion with that individual? These breakdowns in communication are extremely common and while such examples may not constitute scientific misconduct, they certainly encourage fertile ground for contentious relationships and allegation of improper conduct. Indeed, authorship disputes may be one of the most common issues brought to a research integrity office. While very few of these constitute actual scientific misconduct, they nonetheless take time to review and may seriously distort or destroy beginning or long-standing and productive scientific collaborations.

## QUESTIONS FOR DISCUSSION

1. How else might Dr Elliot and Dr Smith have proceeded once they understood Dr Norris' response to their early draft?
2. Is Dr Norris' expectation of senior authorship on all papers from his lab reasonable and/or ethical? What are the pros and cons of using the senior author position to indicate whose laboratory supported the research?
3. Was it appropriate or not for Dr Tatum to accept authorship for assisting with editing the manuscript? What information should Dr Tatum have requested?
4. What is plagiarism? Is the standard used by the institution reasonable? Should it be more specific? Should there be a standard, universally accepted definition?
5. Does publishing the work of a research team but excluding a member of the team who made a significant contribution constitute plagiarism?
6. How might we understand the different conclusions of the inquiry and investigative committee regarding the question of plagiarism?
7. How, if at all, did the possible NIH connection affect the manner in which the RIO processed the case? Was the effect positive or negative?
8. While the ORI's regulations do not apply to "authorship disputes" like the one at issue in this case, institutions are free to prescribe broader definitions of research misconduct. Should universities draft policies prohibiting failure to provide appropriate authorship credit?
9. Could an individual commit research misconduct by demanding authorship credit where their contributions are minimal?

# REFERENCES

Greene, M., 2007. The demise of the lone author. Nature 450 (7173), 1165.

Greenland, P., Fontanarosa, P.B., 2012. Ending honorary authorship. Science 337, 1019.

Martinson, B.C., Anderson, M.S., De Vries, R., 2005. Scientists behaving badly. Nature 435 (7043), 737.

National Academy of Science, 1995. On Being a Scientist: Responsible Conduct in Research. Committee on Science, Engineering, and Public Policy. National Academies Press, Washington, DC.

Price, D.J.S., 1986. Collaboration in an invisible college. In: Little Science, Big Science... and Beyond. Columbia University Press, New York, pp. 119–134.

CHAPTER 9

# Misappropriation and Use of Data from a Multisite and Multi-investigator Study

## 1. PRESENTING COMPLAINT

On January 20, 2007, Dr Dianne Muller of the Department of Dermatology notified her department head and center director of a breach in traditional ethical practice and requested their aid in finding a solution. Dr Andrew Hingston, a collaborating author at another health care institution, had unilaterally submitted an abstract, for a poster presentation at the next annual conference of the American Society of Dermatology (ASD), without first notifying his other coauthors or giving them a chance to comment on the presentation. By doing so, he also failed to disclose important financial information about the study's funding, he violated a contract between the study sponsor and the primary research institution (by not allowing the study sponsor to review a public disclosure of data 30 days prior to disclosure, in order to potentially protect new intellectual property), and he falsely listed himself as the principal investigator (PI) of the study. In addition, Dr Muller asserted that the results of the study had been poorly presented, such that the data analyses were incorrect and led to false interpretation of the research results. After the Chair of Dermatology reviewed these initial complaints, this matter was directed to the Institutional Research Integrity Officer (RIO) for further review and management.

## 2. BACKGROUND AND HISTORY

The study in question was a novel trial of a new monoclonal antibody for treatment of malignant skin melanoma. The final draft of the protocol, by the study sponsor, listed Dr Dianne Muller as the co-PI, and Dr Mark Phillips from the University of Pittsburgh School of Medicine (UPMC, Pittsburgh, PA) was named as the overall PI. The majority of the data were collected at UPMC; with the remainder being generated

The Management of Scientific Integrity within Academic Medical Centers
http://dx.doi.org/10.1016/B978-0-12-405198-0.00009-6
121

mostly by Dr Muller's program and a few cases contributed by Dr Hingston's program.

The study was partially funded by a biopharmaceutical company. The contract between the medical center and the company, dated March 19, 2006, required that any presentation or publication of data from the study must first be provided to the company for review to ensure protection of intellectual property and confidential information, at least 30 days in advance of public disclosure.

## 3. RELEVANT EVIDENCE

In this case, the evidence supporting the initial complaint was never in question. Dr Hingston did submit an abstract to ASD in 2007, which was accepted, and he did present a poster based on this study at the conference. After Dr Muller contacted the Chair of Dermatology, and the problem was made clear to all parties involved, Dr Hingston wrote to the ASD office informing them that the poster had been presented without the knowledge of all coauthors and requested all necessary paperwork to rectify his error. However, as this case investigation proceeded, it became increasingly clear that there was more underlying the case than the single problem of uninformed coauthors.

Dr Dianne Muller, the PI for the local study site, and Dr Phillips, the study's lead investigator at UPMC, were particularly concerned with Dr Hingston's poor analyses and presentation of their data. The entire issue was brought to the attention of the coauthors by an unfortunate circumstance at the scientific conference, in which Dr Hingston presented his poster in a session that directly preceded an oral presentation by Dr Phillips on the very same study results. Dr Hingston had analyzed the data in a way that the experiment was not designed to support, making comparisons that Phillips argued "would be statistically inaccurate and may lead to false conclusions from the data." According to Dr Phillips, even the moderator of the paper session noted discrepancies in Dr Hingston's calculations, placing Dr Phillips and other coauthors in attendance in awkward positions. Of course, both Dr Muller and Dr Phillips were angered by Dr Hingston's use of their data in the first place, without their permission and collaboration. Moreover, on Dr Hingston's poster at the conference, he listed his own hospital affiliation as the lead institution for the study, as well as listing himself as the first author, despite the fact that his institution was involved in collecting data from less than 5% of the participants in the study. Essentially, Dr Hingston's use of data was problematic not only because of his presentation that reflected poorly on his coinvestigators, but because he presented data that did not belong to him, having been collected at other institutions within a clinical trial for which he was not the PI.

## 4. CASE MANAGEMENT AND ADJUDICATION

After receiving the initial presentation of the complaint, the RIO opened an inquiry into the matter. The relevant written evidence consisted of email correspondence and the abstract submitted to the conference (which led to a poster presentation at the meeting). In addition, the RIO conducted interviews with Dr Hingston, Dr Muller, and with the Chief Medical Officer (CMO) for the hospital that employs Dr Hingston. Based on those interviews, it was apparent to the RIO that Dr Hingston had significant difficulty appreciating the root causes for the ire he raised with his colleagues by his actions. Dr Hingston was primarily a practicing dermatologist, rather than a researcher, and he was relatively unfamiliar with the policies and procedures inherent in collaborating on a multi-institutional clinical trial. To complicate matters, the RIO (who is a specialist in aging and dementia) observed that Dr Hingston was exhibiting moderate difficulties with short-term memory and with complex problem solving. As the RIO began to suspect that Dr Hingston—an older gentleman in his early 70s—may be presenting with cognitive impairments (possibly mild cognitive impairment (MCI)), he decided to ask—in confidence—the CMO of Hingston's home institution whether he had any similar concerns. The CMO did confirm this possibility, and noted that they were already considering what, if any, course of action they might pursue from a professional practice standpoint.

## 5. SYNTHESIS AND FORMULATION

From the very beginning, it was clear that there were three main concerns regarding Dr Hingston's actions: (1) the issue of analyzing and presenting study results that were not first reviewed by the study sponsor, per terms of the contractual relationship; (2) the presentation of study results that were factually inaccurate and conflicted with another presentation, by the PI, at the very same conference 1 h later; and (3) the appropriating of property and credit from the rest of the scientific research team. The actual management of this case, however, was complicated by the fact that Dr Hingston was suspected of presenting with a medical/neurological condition that could have limited his capacity to fully appreciate the seriousness of his actions, and the basis for his colleagues' response and complaints.

First, Dr Hingston believed that the data in the study were his to present. The study proposal was closely related to an earlier clinical trial, managed by a regional cooperative research group, for which he had served as a PI. He appeared, to the RIO, to be somewhat confused about the distinction between his prior clinical trial (which had already concluded), and the subsequent, independent launch of this newer trial. While he understood that Dr Muller was named as PI for the full protocol, he assumed that the study itself was a continuation of his own research and that as such he was free to present the data as he wished. Dr Hingston also placed most of the blame for his coauthor's lack of knowledge of his presentation on a communication failure by his secretary and his assumption that ASD had contacted his coauthors, whose email addresses he had provided. Of course it is the case that, as the individual proposing a presentation for an annual conference, it was his personal responsibility to ensure that all of the proper procedures and notifications had occurred.

Second, Dr Hingston analyzed the data in a way for which the experiment was not designed. While his coauthors were upset that his representation of the study's initial results were inaccurate and poorly presented, Dr Hingston did not seem to understand their concerns. In contrast, Dr Phillips, the lead investigator for the study at UPMC, reported that Dr Hingston presented a "warped sense of the data" and noted that "even the moderator of the session noted the discrepancy" in his analysis. While this is a secondary issue, considering that even a proper analysis of the data would have been a breach of contract and faith between Dr Hingston and his coinvestigators, the poor quality of Dr Hingston's analysis only exacerbated his coauthors' anger.

Finally, Dr Hingston was apparently unaware of the financial and legal contracts between BMS and the medical center regarding the support of this clinical trial, and the requirement to allow the sponsor to review any presentation or publication in advance of public disclosure. With proper communication between Dr Hingston and Dr Muller, this issue could have been entirely avoided.

## 6. RESOLUTION

The RIO concluded this investigation and summarized his decisions in a letter to the involved parties on February 10, 2007. His main recommendations were as follows:

1.  *Withdraw the abstract from the ASD conference web site*
    Based on the initial complaint, Dr Hingston had contacted the ASD office to request a new disclosure form, to be signed by all coauthors.

Based on the seriousness of the breach in procedure, contract, and research integrity that Dr Hingston's original submission caused, this action was not sufficient to remedy the damage. The RIO insisted that Dr Hingston withdraw the abstract completely and request that it be removed from the ASD conference web site. In addition, the study's coauthors submitted a second abstract to another conference specifically to correct the misinformation contained in the ASD conference poster the preceding year.

2. *Ban Dr Hingston from presenting any more abstracts based on work with the academic medical center, or from participating in any future studies with the medical center or its affiliates*

In order to limit any similar breaches of contract and procedure in the future, Dr Hingston was also asked to refrain from submitting any abstracts, presentations, or papers based on the study in question. This step was necessary because Dr Hingston did not seem to understand the magnitude of his mistake or the root issue (publishing data belonging to others, without permission), instead focusing on a simple communication failure as the cause of the problem. For similar reasons, the RIO recommended that Dr Hingston be barred from working as the principal- or coinvestigator on any future collaborations with coauthors from this medical center or its institutional affiliates (this ban, of course, does not include the hospital that employs Dr Hingston, which is unaffiliated with the academic medical center).

3. *Recommend that the medical school reevaluate Dr Hingston's faculty position*

The final recommendation followed from the questions raised by the investigation regarding Dr Hingston's cognitive status. According to interviews conducted by the RIO, administrators at Dr Hingston's hospital had previously noticed the same cognitive difficulties emerging over the past 1–2 years, possibly as a result of advancing age and/or incipient disease. Whereas that hospital had not yet taken any action to seek a formal assessment, the RIO agreed with the medical school dean's office that his clinical faculty appointment would be allowed to (quietly) lapse when his current term on faculty expired in approximately 6 months. At the time when the current term expires, the medical school agreed to notify Dr Hingston of the decision to seek his voluntary retirement from the faculty.

## 7. COMMENTARY

This case raised questions regarding authorship and data ownership, and how clear evidence of research misconduct might be considered, when there is reasonable suspicion that the individual in the center of the controversy may be suffering from a condition that limits his or her ability to fully appreciate one's personal conduct.

## 7.1  Use of Collected Data

As noted above, the main complaint that Dr Hingston's coauthors had with his early presentation of material at ASD was not merely that it was presented inaccurately (with regard to statistical treatment of the data), but that it was presented without their knowledge at all, and that Dr Hingston grossly misrepresented his own role on the project. When Dr Hingston presented his initial analysis of the study, he not only listed himself as the PI, but he listed his employer as the primary research institution. Dr Muller and Dr Phillips contended that Dr Hingston was not only responsible for misrepresenting his role in the investigation, but also for analyzing and publishing data that did not belong to him, essentially stealing their property.

The question of ownership of data is difficult to parse, particularly in this age of digital collaboration between multiple institutions on large-scale research projects. Dr Hingston was originally included as a coinvestigator on the study because of his care of patients who might benefit from participation in the trial, but he may never have fully understood the subtleties of the protocol design or statistical approaches to proper data analyses. Even for experienced researchers, precise details regarding data ownership may be unclear. Culliton (1988) noted that "ownership of research data has, until recently, never been much of an issue," because scientists generally agreed that data belonged to the individual researcher(s). Others have argued that "the institution, not the investigator, is probably the ultimate owner" (Culliton, 1988). With large multisite trials, for which data sets are collected by a number of researchers and pooled for analyses, who owns these pooled databases? Whether the researcher who collected the majority of individual data points, the scientist listed as PI on the protocol, or is everyone partially responsible for its ownership? A lack of consensus remains, with respect to the creation and ownership of research data especially in multisite, multi-PI research efforts. Josh and Krag (2010) outlined several common concerns, including the lack of a precise definition for "data," agreement over standard storage methods and times, and responsible data sharing. Dr Hingston's collaborators clearly believed that not all coinvestigators should enjoy equal rights to the use of the pooled study data. If Dr Hingston had properly sought permission from his collaborators, and expressed his interest in presenting study results at the ASD conference, would he have been allowed to do so?

## 7.2  Consideration of Medical Complications

In this case study, Dr Hingston, the investigator who submitted the offending abstract, had enjoyed a very long career as a physician and as Chief of Dermatology at his hospital. He also held the post of Clinical Associate Professor at the affiliated medical school. This research integrity inquiry served to raise questions regarding his cognitive competencies as both an investigator and as a practicing clinician. The RIO discussed his own observations and concerns (after his initial interview with Dr Hingston) with the CMO of the hospital that employed Hingston, and it became clear that the hospital leadership had similar concerns that Dr Hingston may be suffering from early stages of a dementing illness. As such, the RIO took this into account as an extenuating circumstance (see above). The entire issue of how best to manage older, well-respected professionals who may be suffering from age-related and/or disease-related changes in cognition is very difficult.

In this case, the RIO opted for a more moderate response, which would allow Dr Hingston to end his research career, and his association with the medical school, with dignity. Dr Hingston's contract with the medical school, as a professor, was up for renewal within a few months of the incident; and a decision was made to simply decline to renew this agreement (medical school policies do allow for the individual faculty member to appeal such decisions). In addition, on the RIO's recommendation, the institution's Chief Research Officer barred Dr Hingston from further research activities within the medical center, and this was done quietly and without drawing any attention to this decision (unless or until Dr Hingston might attempt to engage in future research activities). With these actions, Hingston's long career was not marred by a widely visible ethical breach, just prior to his own retirement.

Dr Hingston's clinical position, at his local hospital, was a matter that is not under the purview of the RIO. Rather, the hospital's clinical leadership is responsible for the evaluation of clinicians with suspected medical (or other) issues that might impair judgment. Having noticed Dr Hingston's confusion during conversations, apparent difficulties with short-term memory, and diminished problem-solving skills, it is appropriate for a formal neuropsychological examination to be conducted, with the goal of ensuring that any risk of inadvertent harm to patients is minimized, and at the same time striving to preserve his dignity as an older professional.

## QUESTIONS FOR DISCUSSION

*Parties*: Dr Andrew Hingston, defendant, collaborating author for study; Dr Dianne Muller, co-PI for study, complainant; Dr Mark Phillips, overall PI for study.

1. In this case, the RIO weighted the fact that Dr Hingston was very close to retirement after a long career, and that he was beginning to show evidence of MCI, very heavily in making recommendations to the deciding official and Chief Research Officer for the medical center. Would you have done the same? To what extent should such individual issues influence the outcome of such a case, where the actual misconduct is not in question?

2. Was the RIO correct in asking for a confidential conversation with the CMO of a hospital outside of his or her own system? The RIO discussed Dr Hingston's cognitive status with the CMO of Hingston's employer. Was this acceptable, given the circumstances and concerns?

3. Rather than sending a strongly worded letter to Dr Hingston, banning him from any future collaborative research with faculty employed by the RIO's academic medical center, the RIO and Chief Research Officer instead decided to only enforce this decision in the unlikely event that Dr Hingston were to initiate a new research project at the medical center (which did not occur prior to his retirement, several months later). Was this a mistake? Why?

4. Was the RIO's behavior affected by the possibility of a breach of contract suit by the study sponsor?

5. Dr Hingston does not appear to have committed research misconduct, as defined by Federal regulations (fabrication, falsification, or plagiarism). In light of this fact, how could the RIO sanction him? Did the institution have a broader definition of research misconduct (perhaps including misattribution of authorship).

6. Who "owns" data? Whether the researcher, the institution, or the Federal government (if the research is federally funded)?

7. Was it appropriate for the RIO to treat Dr Hingston's possible illness as an extenuating circumstance?

## REFERENCES

Culliton, B.J., Nov. n, 1988. Authorship, data ownership examined. Science 242 (4879), New Series.
Joshi, M., Krag, S., 2010. Issues in data management. Sci Eng Ethics 16, 743–748 (Web).

## CHAPTER 10

# Determining the Extent of Data Fabrication Following an Apparent Single Incident

## 1. PRESENTING COMPLAINT

In the late spring of 2010, Dr Taylor, a junior pathologist, spoke with her clinical lab director and mentor, Dr Owens, and confessed to having misrepresented work both to Dr Owens and in a clinical research meeting presentation. Specifically, she admitted to using a different immunohisto-chemical staining for human kallikrein in over 100 samples of ovarian cancer from a sample repository than those she was supposed to have used. Further, in lab meetings, in conversations with Dr Owens, and in a manuscript in preparation to be submitted, she had misrepresented the results of protein activity in the tumors as if these reflected the immunohistochemical staining she was supposed to have used rather than the process she did use. She had also included some of the data in question in an abstract and oral presentation at a regional academic meeting. Importantly, save for the meeting abstract, none of the findings had been published. While the work did not involve NIH funding, Dr Taylor's postdoctoral clinical research fellowship was supported by an NIH training grant.

## 2. BACKGROUND AND HISTORY

Dr Taylor had joined her current lab in pathology two years ago following her residency as a pathologist specializing in oncology. She was especially interested in immunohistochemical markers as predictors of prognosis in ovarian cancer and had joined her current lab as part of a physician scientist training program with a specialization in immunopathology and oncology. While her work with Dr Owens progressed slowly given related clinical demands on an oncology service, she was still able to complete several experiments with tissue samples from patients with ovarian cancer and contribute her findings for a regional research meeting.

*The Management of Scientific Integrity within Academic Medical Centers*
http://dx.doi.org/10.1016/B978-0-12-405198-0.00010-2

Before joining Dr Owens' laboratory and during her pathology residency, Dr Taylor had also been able to work in a productive pathology laboratory in another institution and, working with a mentor, she had completed two experiments and had published two papers in moderate-impact clinical pathology journals. Dr Taylor had been eager to join Dr Owens' laboratory to continue lines of work synergistic with those she started during her residency. She had come to her current institution just after her marriage and soon she was pregnant with her second child. Her first child had been born in the last year of her residency and was doing well. But within the first year of beginning her postdoctoral clinical research fellowship, she was again a new mother with added responsibilities at home as well as a growing portfolio of experiments in the laboratory and responsibilities as a clinical pathologist two days a week. Her husband was an IT specialist in a new biotech company and had an equally demanding, time-consuming job. Quiet but hard working, Dr Taylor was determined to move her career as a physician scientist forward.

Dr Owens' laboratory had a reputation of being supportive to young physician scientists and encouraging of their career development. Relatively young in his own career though a full professor, he had been able to maintain an atmosphere conducive to discussion and was readily available to his team even as he was working especially hard to maintain steady funding support for the lab. Dr Taylor's admission was startling and disconcerting especially as Dr Owens was so careful to encourage open communication with this fellows and lab staff. He had difficulty imagining why Dr Taylor would not have simply told him she could not process and review the samples as they had discussed for whatever the reasons might be. He had always thought Dr Taylor to be honest and conscientious.

Indeed, what Dr Owens did not know was that Dr Taylor's second child was seriously ill and her husband had recently been laid off from work when his biotech company had been forced to downsize with the unstable economy. Their respective families were far away and they had few to no friends to help either with their first child or with the growing number of medical appointments. They were isolated, under considerable stress, and Dr Taylor was well aware that soon she would need to be looking for jobs as her salary was the sole economic support for their family and her work-related health insurance was now their only source of support for their infant's medical needs. While she knew she was able to find a job as a clinical pathologist in many settings, she was very focused on establishing a clinical research career and wanted a job as a junior faculty blending research with her clinical responsibilities in a specialized pathology service. But with the disruption of moving, her new family, the birth of her second child, and her clinical responsibilities, Dr Taylor's first two years with Dr Owens had not been as productive as she had hoped. While she had learned new techniques and processed a large number of samples for new genetic markers, no manuscripts had been submitted for review, and she had presented only once at a regional conference. Even with her quiet determination, her postdoctoral fellowship was not going as well as she had planned and her young family was certainly struggling under considerable stress.

## 3. RELEVANT EVIDENCE

After Dr Taylor spoke with Dr Owens about her failure to accomplish the assays as designed and her apparent intentional deception of both her Personal Investigator and other colleagues in the lab, Dr Owens sought consultation from the Research Integrity Officer (RIO) and university attorneys. He provided a letter summarizing his conversation with Dr Taylor, the abstract that Dr Taylor had submitted, and a draft of the manuscript in preparation. At this initial point in the investigation, these three pieces of information covered the relevant evidence in the matter.

## 4. CASE MANAGEMENT AND ADJUDICATION

Based on consultation with the university attorneys and Dr Owens' concerns, Dr Taylor's postdoctoral fellowship appointment was terminated for cause based on her admission of lying to Dr Owens and her apparent effort to deceive both her immediate colleagues in the lab and the scientific community through her presentation at a regional conference. She was also removed from her clinical responsibilities because of the questions raised about her integrity. Because of the NIH support for Dr Taylor's fellowship, the RIO convened an inquiry committee charged with the task of examining Dr Taylor's work that she admitted to falsifying. The inquiry committee was charged with reviewing Dr Taylor's voluntary admission to Dr Owens, the impact on the rest of the work she was involved in, and whether or not an investigation was warranted into the other areas of work Dr Taylor was involved in.

The inquiry committee took care to interview those administrative officials involved in Dr Taylor's termination to assess as much as possible Dr Taylor's state of mind and ensure themselves that her admission was completely voluntary. They learned that Dr Taylor did not protest the termination of her appointment though she was very concerned about her clinical responsibilities when she was given the letter describing her termination and was escorted back to both her research lab and her clinical pathology workstation to retrieve her personal belongings. The inquiry committee also interviewed Dr Owens and Dr Taylor. Dr Owens reported that he was beginning to check other experiments and samples Dr Taylor had contributed to. Dr Taylor also acknowledged to the committee the same data falsification she had voluntarily described to Dr Owens. The committee inquired of how Dr Taylor had gotten to the point of falsifying data. Dr Taylor admitted to feeling increasingly unable to keep up with both her research and her clinical responsibilities and being under considerable stress and feeling desperate. She felt depressed and thought she would buy time with presenting results from "easier" experiments with every intention of at some point actually doing the experiments she and Dr Owens had discussed. But time alluded her and knowing that she had intentionally misrepresented her work only created more pressure and worry as she fell further and further behind in her self-appointed career goals. When asked specifically, she confirmed again that she had intentionally used immunohistochemical staining for human kallikrein different from the one she and Dr Owens had discussed and that she had presented findings from those assays as if these reflected the experiment as designed and described. The inquiry committee recommended a more detailed investigation into whether or not Dr Taylor had committed any other acts of data manipulation or falsification in her other work in Dr Owens' lab. Dr Taylor was not forthcoming in this regard though, given her state of mind, it was difficult to know for certain how extensive her apparent scientific misconduct may have been.

Based on the inquiry committee's recommendations, a more detailed investigation was undertaken about all of Dr Taylor's work in Dr Owens' laboratory. At this stage in the process, the institution was also keeping the NIH Office of Research Integrity (ORI) informed of the investigative steps given that Dr Taylor's work was supported by NIH training funds. As part of the investigation, the committee again spoke at length with Dr Owens and Dr Taylor and examined Dr Taylor's lab notebooks and her computer files. The investigation attempted to verify the assay as described by Dr Taylor by hiring an independent laboratory technician not associated with Dr Owens' laboratory to examine the samples of ovarian tissue as prepared by Dr Taylor and to repeat the immunohistochemical staining in a select sample of the frozen tissue using the originally intended human kallikrein immunohistochemical staining as well as the one Dr Taylor said she used. Careful study of these three sets of processed tissue revealed a further unexpected finding. The set of tissue processed in the way Dr Taylor said she had done did not correspond to the already processed samples Dr Taylor provided the committee. Her work appeared both different from what she and Dr Owens discussed and, through the attempt to replicate the assay she said she used, also different from the results she reported. Thus, while it was confirmed that Dr Taylor had presented false data based on the intended but not actually used immunohistochemical staining, it also appeared that in her voluntary admission, she had also incorrectly and perhaps intentionally not revealed other variations in the tissue processing. In terms of other experiments in the lab where Dr Taylor played a minor role, data were independently confirmed and/or replicated.

Because Dr Taylor's postdoctoral fellowship was supported by NIH training funding, the ORI was notified of Dr Taylor's voluntary admission and of the institution's inquiry and investigative process. University officials regularly informed ORI officials about the committee's findings and the stages of inquiry and investigation. Based on the thoroughness of the committee's work at each stage of the process, ORI accepted the university's findings and recommendations and did not institute an independent investigation.

## 5. SYNTHESIS AND FORMULATION

This case is unusual because Dr Taylor voluntarily came forward to her mentor and employer, Dr Owens, and admitted her actions. Her admission immediately entailed both a reporting to the RIO and involvement of counsel around termination of Dr Taylor's employment for cause. The additional extensive investigation of other aspects of Dr Taylor's work in the lab was on the one hand reassuring in revealing her falsification was apparently limited to the experiments she voluntarily described.

On the other hand, the investigation revealed more confusing evidence suggesting that Dr Taylor's voluntary admission of her falsification may not have conveyed the full story of her apparently intentional deception. Hence, while Dr Owens felt that Dr Taylor's falsification was limited to the work she did on the experiments in question, it was still not clear that Dr Taylor was completely candid in her description of the extent of her data falsification. It may well be that she was having such difficulty with her work at that point that her own recollection of her falsification actions was faulty. Alternatively, it is also possible that she was knowingly deceiving Dr Owens even in her voluntary admission though the motive for such a deception would be less clear. Nonetheless, the investigation did clearly confirm that Dr Taylor had falsified data in claiming to run experiments different from those she had actually completed. At the same time, what seemed clear is that with her husband's unemployment, their infant's serious medical issues, her clinical responsibilities, and the family's social isolation, Dr Taylor herself became seriously compromised in her ability to work effectively. Her decisions were impaired and confused and once she made the first step toward deception in not running the experiment as planned but presenting the results as if she had, she was in a trap that spiraled into further deception and data falsification. Fortunately, she was able to find a way to tell Dr Owens, but it was not clear that Dr Taylor herself really understood what had happened and how she might care for herself, learn from the incident, and not find herself in a similar situation in the future.

## 6. RESOLUTION

As described above, Dr Taylor's fellowship was immediately terminated following her voluntary admission of her actions. She was able to obtain another job as a clinical pathologist not involved in research in a city closer to extended family and quickly relocated with her husband and their infant. Her husband was able to do some part-time IT consulting work and stay at home to take care of their infant's medical needs as well as transport their older child to and from preschool. The investigation committee advised Dr Owens to inquire about the possibility of withdrawing the regional conference abstract that appeared to contain the falsified data. Inquiries into this possibility revealed it to be difficult to remove abstracts from conference proceedings that are not disseminated beyond the conference. The investigation committee also recommended that the administration notify Dr Taylor's new employer about her voluntary admission of data falsification and the findings of the investigation. After consultation with the university counsel, the decision was made to notify Dr Taylor's new employer. Dr Taylor was notified of this action and her new employer, the director of a large clinical pathology group, acknowledged receipt of the notification.

## 7. COMMENTARY

This case raises a number of points for discussion. First are the procedural issues for addressing scientific misconduct in instances of voluntary admission. While initially, voluntary admission to scientific misconduct may seem straightforward, albeit tragic, it nonetheless raises the question of how extensive is the individual's wrongdoing. Does it predate current lab employment? Does it extend to earlier training periods? Indeed, while on the one hand a voluntary admission speaks to personal integrity, it also potentially calls into doubt all aspects of the individual's work, especially because the scientific enterprise is so dependent on individual honesty and openness to unexpected findings, failed experiments, and the clarifying force of scientific debate. Even a voluntary admission may, or should, require careful review of the individual's entire scientific portfolio. There are many examples of discoveries of fraudulent scientific activities being only one of a long history of misconduct often going back years and across different laboratories (Fanelli, 2009; Marshall, 2000; Martinson et al., 2005). Thus, a voluntary admission does not make less work for an investigative process. It only jump starts the process with the initial cooperation of a respondent.

The present example also shows how, sadly, once an individual starts down a path of deception, it can be difficult to recall or even track all the steps in that deception. That efforts to confirm the details of Dr Taylor's deception revealed discrepancies from what she admitted to only raises more concerns about her integrity and ability to honestly discuss her scientific efforts. The surprise in this investigation suggests that even with voluntary admission, it may be important for an institution to attempt to determine, if possible, the verity of the admission and if the falsified procedure or data appear to be distorted in the ways described by the individual.

A third point raised by this case is how to institutionally respond to discovered versus voluntary admission of wrongdoing. Surely, it is difficult for any laboratory or scientific group to maintain someone on the team who is no longer viewed as trustworthy no matter how forthcoming in their admissions they have been. The overall health of a lab and institutional research enterprise depends on trust and respect among scientists and scholars at all levels of experience. In this instance, that concern may have been heightened even more because of Dr Taylor's involvement in clinical pathology with the responsibility of reviewing many patient tissue samples. Her clinical integrity impacted treatment decisions by the oncology team. At the same time, a voluntary admission of wrongdoing suggests an individual under considerable psychological distress who is at a moment when he/she may most need the support and help of familiar colleagues and mentors. For some individuals, support and an opportunity to restore their reputation may help them understand their own motivations in their actions and help prevent further intention or other-wise errors in judgment in the future. There is also the concern that an individual under such duress that impairs judgment and integrity may move to another institution without fully working through their difficul-ties and thus continue their misconduct, an action that threatens not only the overall integrity of the scientific (and clinical) enterprise but also potentially impacts interinstitutional relationships. Hence, there are always considerations of when and how to inform other institutions without inappropriately impacting the future opportunities for the individual to repair and restore his/her career. There are no single appropriate or cor-rect approaches to these dilemmas save for encouraging open, frank dis-cussion about the best course of action given the circumstances and the individual in question.

Finally, perhaps most poignant if not generally illustrative is the cir-cumstances that seemed to lead to Dr Taylor's actions. Depression, stress, social isolation, and a heavy workload appear to have conspired to impair Dr Taylor's judgment and one serious misstep led to another and then another. Surely, not every young physician scientist struggling with fam-ily or personal issues commits scientific misconduct and most often, the personal circumstances and motivations leading individuals to engage in scientific misconduct are never totally clear (Davis et al., 2007; Neaves, 2012). It should also be stressed that we only know at the surface those conditions that impaired Dr Taylor's judgment. It may well be even she was not able to reflect more deeply of what led to her deceptive actions. At the same time, these circumstances burdening Dr Taylor and many at her stage of development set up a tragically fertile ground for miscon-duct, especially when combined with the pressure to be productive and build a portfolio of work to support future job applications. Dr Taylor's example is especially tragic in her inability to reach out and tell Dr Owens or some other trusted colleague about her increasing psycho-logical distress and the considerable stressors in her family. It is especially true in very intense, pressured, and productive environments, individuals in psychological distress may quickly fall behind in their work which only intensifies their stress and impairs their judgment. Every institution has employee health resources ready to assist individuals in distress but helping young scientists such as Dr Taylor access those services requires both knowing about their distress and also having the skill and tact to present such services as health promoting rather than indicative of failure.

## QUESTIONS FOR DISCUSSION

1. Should Dr Taylor have been immediately terminated from her fellowship? Might there be other options to respond to her voluntary admission of falsification?
2. Discuss the decision to notify Dr Taylor's new institution. What are the extenuating circumstances that may influence such a decision positively or negatively?
3. Do the findings of the more extensive investigation process alter the initial impression of Dr Taylor's voluntary admission? If so, how?
4. Might there have been warning signs regarding Dr Taylor's increasing stress and inability to work? In retrospect, are there preventative actions that may have been helpful?
5. Why were the university's attorneys brought into the discussion at such an early stage? Is this something that should typically occur when allegations of research misconduct are raised?
6. Should the institution have considered proceeding with an inquiry before terminating Dr Taylor's fellowship for cause? Or did her admission of misconduct obviate the need for this formal process?
7. What are the implications of the decision to notify Dr Taylor's new employer of her research misconduct? Is there any risk that this notification may have risked the possibility of a tortious interference action? Why or why not? What arguments would the institution make against such a claim?

## REFERENCES

Davis, M., Riske-Morris, M., Diaz, S., 2007. Causal factors implicated in research misconduct: evidence from ORI case files. Sci Eng Ethics 13 (4), 395–414.

Fanelli, D., 2009. How many scientists fabricate and falsify research? a systematic review and meta-analysis of survey data. PLoS ONE 4 (5), e5738.

Marshall, E., 2000. Scientific misconduct – how prevalent is fraud? That's a million-dollar question. Science 290, 1662–1663.

Martinson, B.C., Anderson, M.S., de Vries, R., 2005. Scientists behaving badly. Nature 435, 737–738.

Neaves, W., 2012. The roots of scientific misconduct. Nature 488, 121–122.

CHAPTER 11

# Anonymous Allegations of Scientific Misconduct

## 1. PRESENTING COMPLAINT

Shortly after a high-impact paper appeared in *Lancet*, the institution of the senior author received an anonymous allegation by e-mail. The complainant, identified only by initials and e-mailing from a presumably anonymized e-mail address, made allegations about one aspect, the "behavioral data," of a complex paper on the neurobiology of learning and aging in an animal model. The complainant alleged the data presented were not only inconsistent with the investigators' interpretation but also were not replicable. The complainant further alleged that the data suggested deliberate falsification over a number of years. The research in the paper was supported by NIH funding. Further, the first author, Ms Fields, was a very talented undergraduate student doing an intensive senior year of research in Dr Edward's laboratory. Within a few weeks of publication she was presenting her senior thesis for graduation that was based largely on data in the published paper. The complainant's allegations primarily targeted the first author's work.

## 2. BACKGROUND AND HISTORY

Dr Edwards' laboratory was internationally recognized for programmatic research on cognition and learning with aging. Dr Edwards' research team had made seminal contributions on learning in the context of aging and on the changes in learning with advanced age at the cellular, genetic, and functional level. Dr Edwards successfully collaborated with many labs within her own institution as well as outside her institution. In particular, she was able to examine research questions at several complementary levels of analysis by collaborating with other labs with expertise in relevant methods such as cellular electrophysiology, epigenetics, and drug development. The *Lancet* paper in question involved three laboratories, two in the same institution and another outside but each contributing data to the paper and each involving both senior and

*The Management of Scientific Integrity within Academic Medical Centers*
http://dx.doi.org/10.1016/B978-0-12-405198-0.00011-4

junior investigators. The lab led by Dr Barnes was responsible for detailed cellular electrophysiology recordings in two experimental groups of animals.

The first author, Ms Fields, had joined Dr Edwards' laboratory as a graduate student three years earlier. She had spent a number of years working as a research assistant and summer intern before beginning her senior thesis research in the summer before her senior year. She was unusually gifted and brought an already impressive amount of laboratory experience to her undergraduate major and senior thesis work. She had designed her thesis around a series of experiments investigating synaptic changes in older animals and she was targeting a specific neural mechanism for learning. Her thesis research had involved a number of experiments leading up to a year of work for the large study reported in the *Lancet* paper. She had first authored another paper in moderate-impact journals and was a co-author on two other papers, an impressive portfolio for someone at the beginning of her research career. Appreciated as hard working, collaborative, and careful, her lab team was very supportive and knowledgeable of Ms Fields' work. The studies reported in the *Lancet* paper involved many animals with multiple behavioral tests and observations in addition to studies of synaptic morphology.

## 3. RELEVANT EVIDENCE

On receipt of the allegation by e-mail, the Research Integrity Officer (RIO) sought the advice of two senior faculty with sufficient expertise in the relevant fields covered by the manuscript to evaluate the specificity of the allegation. Additionally, the RIO sent an e-mail to the anonymous complainant acknowledging receipt of the allegations and asking if the complainant(s) could be more specific about which data in the paper they were alleging were falsified or fabricated. The complainant responded within a short time pointing to specific figures and revealed more specific knowledge about the students in the lab, as well as the fact that he/she had had access to at least some of the raw data and behavioral recordings and that they had been involved in unsuccessful efforts to replicate aspects of the synaptic morphology experiments. With greater specificity from the complainant and direct allegations of fabrication, the senior faculty advisors recommended moving to an investigation of the allegations.

## 4. CASE MANAGEMENT AND ADJUDICATION

Based on the recommendation to move to an investigation process, the RIO convened a committee with relevant expertise and made plans to notify Dr Edwards' lab, including Ms Fields. It was critical to protect relevant evidence at the same time as notifying the laboratory team and thus, at the time of notification, the RIO, working with legal counsel, sequestered all relevant data. To do so, the RIO and legal colleagues met with all authors on the manuscript who were gathered together by the senior corresponding author, Dr Edwards, within an hour of his being notified by the RIO of the allegations. In the course of meeting with the eight co-authors, the RIO identified all relevant data that needed to be secured immediately in the presence of the eight authors. This included copying the hard drives of the experiment computers, the hard drives of any computers including personal laptops used for data analysis and/or manuscript preparation, securing e-mail records on the university server, and securing raw data including PCR films, animal cage records, and sign-in files for the animal facilities. This also included the raw data from the cellular electrophysiology recordings made by the collaborative laboratory. The sequestration process may have been quite stressful to laboratories inasmuch as the team members, including those from the collaborative laboratory, were not allowed to leave until the RIO had secured all relevant data. At the time of notification by the RIO of the allegation, Dr Edwards' team tried to understand where the anonymous complaint had come from and speculated about who may have been sufficiently knowledgeable to make such specific allegations.

The faculty committee interviewed all members of the research team including Dr Edwards and Dr Barnes and Ms Fields. They also offered the anonymous complainant the opportunity to speak with the committee either by an e-mail or chat conversation or via phone or Skype. The complainant declined to speak with the committee.

## 5. SYNTHESIS AND FORMULATION

Each member of the research team involved in the manuscript worked on specific, overlapping aspects of the data. Dr Barnes and his team were responsible for the cellular electrophysiology as well as the synaptic morphology studies. The figure with the morphology findings had been called into question by the complainant. Dr Barnes provided the faculty committee with the original data used to generate the relevant figure and described in detail how the morphology measurements were accomplished. The faculty committee had sufficient expertise to review the raw data and morphology measurements independently and concurred with the findings reported by Dr Barnes and his team. Dr Barnes also reported that he and his team had replicated the findings reported in the paper in another set of experiments done independently from Dr Edwards' laboratory.

Other members of Dr Edwards' laboratory expressed their surprise and confusion because of the allegations. While Ms Fields had performed all the behavioral studies on her own, as was customary in Dr Edwards' and other laboratories, the senior research staff in the lab independently spoke about her care with data and their confidence in her approach to the behavioral data. She had been trained by the senior research associate and lab manager and had consistently been reliable with both in the training sessions.

However, because of the emphasis in the anonymous complaint on the behavioral data from the animals, the faculty committee recommended hiring an independent expert to rescore animal responses to the learning cues based on recorded sessions available from the experimental computer. A person was engaged from another institution with considerable experience in the paradigms in question. This individual rescored from the video recording of all the experimental sessions with the animals and obtained comparable data to those in Ms Fields' raw data files and to the data reported in the figures in question.

## 6. RESOLUTION

Based on the faculty committee's thorough review, the decision was made not to proceed with further investigation. The committee felt that Ms Fields' approach to the behavioral data recording and scoring was consistent with the standards of the field and that independent verification of her scoring had found consistent data to those she reported. Further, raw data provided by the collaborating labs was consistent with that reported in the manuscript. Thus, there appeared to be no reason to believe that the allegations were true and no reason to proceed with a more detailed investigation. Ms Fields successfully presented her senior thesis, graduated with honors, and proceeded to her graduate studies at a prestigious international graduate program. The anonymous complainant was informed of the faculty committee's findings and did not respond further. ORI was also informed of the findings of the inquiry process and accepted the institution's process and recommendations.

## 7. COMMENTARY

The central feature of this case is the anonymous complaint that triggered the inquiry process. It is increasingly easy for individuals or groups of individuals to file anonymous complaints through e-mail. Anonymous complaints present special challenges for any research integrity oversight. For one, depending on how willing the complainant is to participate, it may still be difficult to obtain clarification or additional information from the complainant. Second, it is especially difficult to evaluate whether or not the complaint was brought in good faith without knowing some details about the complainant. That is, on some occasions, complainants bring allegations out of anger or revenge toward a collaborator or senior colleague because of disagreements either around the research in question or past projects. Anonymously submitted allegations do not give inquiry committees the opportunity to evaluate even superficially the nature of the relationship between the complainant and respondent and to determine whether or not there are relevant conflicts of interest coloring or even driving the allegation. For this reason, RIOs and journal editors are often reluctant to act on anonymous complaints or, at the very least, to insist that the anonymous complainant provides some documentation as to their identity even if that identity is not revealed to the respondents (Editors' Update, 2013; Grens, 2013).

However, it does appear that anonymous allegations to journals and universities may be increasing (Oransky, 2013; http://retractionwatch.com/) as is in general the number of allegations. The number of retractions of published manuscripts also seems to be on the increase. In 2011, there were 381 journal retraction notices, an increase of 22 compared to 2001 as tracked by the Thomson Reuters database Web of Knowledge (Gewin, 2012). Hence it is important that the scientific community develops guidelines for how and when to respond to anonymous complainants as well as complaints in general. For this reason, the Committee on Publication Ethics (COPE; http://publicationethics.org/about/council/virginia-barbour) has responded to journal editors' concerns about the increasing volume of anonymous complainants with some guidelines. Fundamentally, the recommendation from COPE is to take any potentially legitimate complaint seriously regardless of whether or not the individual discloses his or her identity. The challenge remains in determining legitimacy. In the present example, the complainants provided very specific details about their initially broad allegations suggesting at the least a close read and consideration of the publication and at the most, first-hand knowledge and/or opinions about the research team. Again, the challenge is to sort through potential conflicts of interests and personal grudges.

A second implication of the present case example is the impact on very junior scientists just beginning their careers and especially when they are at critical stages in their early research and publication histories. While exposure to data falsification and plagiarism is less common, it is nonetheless surprisingly common for undergraduates and graduate students to encounter other types of misconduct and questionable research practices (Swazey et al., 1993). A faculty mentor's scientific behavior and standards may have a significant influence on the formation of a student's values and standards; and how these matters are handled at the institutional level may also shape a student's approach to both scientific integrity and to coming forward if he/she witnesses questionable practices. Students' opinions and attitudes are also shaped by how senior faculty responds when they find themselves as respondents in such matters. In the present example, all members of the research team were immediately forthcoming and Dr Edwards carefully kept her team focused on helping the faculty committee rather than focusing on their anger and dismay at the anonymous complainant. The forthcoming, cooperative, and transparent approach Dr Edwards modeled surely made an impression on all members of her research team. Far more difficult to assess is the impact of an allegation and inquiry process on a student's attitudes toward a scientific career, in general, even if they are found innocent of any wrongdoing. These processes are stressful and difficult for all scientists at any level of experience but may be especially formative for students just beginning their careers and shaping their attitudes toward the scientific enterprise as a whole. Support for students in these positions is critical with follow-up by a trusted mentor and colleague even well after the inquiry process is complete.

Finally, as science becomes increasingly inter- and trans-disciplinary, research necessarily involves collaborating labs from multiple disciplines, publications often involve data collected by separate labs within and across institutions, and research teams integrated into one comprehensive account of the particular research question. Indeed, multidisciplinary teams may be quite productive and innovative (Disis and Slattery, 2010; Porac et al., 2004). At the same time, the question becomes does every author then have the responsibility to fully speak for all data, even those outside their expertise? While the senior or first and corresponding author may still retain primary responsibility for all data and findings in a manuscript, it is also now not uncommon for there to be two or more shared senior or first authors representing the collaborative efforts. It is also difficult as science becomes more specialized as well as necessarily more collaborative for any one scientist to be able to fully evaluate the data from another highly specialized collaborator. While allegations may target just one aspect of a publication or scientific report, it is common practice then to interview and evaluate all contributors to the manuscript or project in question. Not only does this increase the time required for the inquiry process but also spreads the burden and stress of the process across laboratories. Additionally, the more laboratories involved, the more difficult it is to maintain full confidentiality. While there are no simple answers to these kinds of concerns and trends, it is important that committees and research integrity officers are cognizant of the increased spread of any allegation when there are multiple laboratories contributing to the program of research.

## QUESTIONS FOR DISCUSSION

1. Discuss the approach to anonymous allegations and how it is or is not possible to evaluate fully the legitimacy of the complaint. Should an institution have a policy on either not evaluating any anonymous complaint or of proceeding forward on all? Should there be criteria for proceeding? Should an anonymous complainant be required or at least strongly urged to reveal his/her identity? Should the anonymous complaint have been pursued in the present example? Should the identity of the complainant have been pursued?

2. Can you think of any effective and potentially innovative ways to better educate students about proper scientific practice and how to recognize, and come forward, around questionable scientific practices? Similarly, how might students involved either as a primary respondent or part of a research team be mentored and supported through an inquiry/investigative process?

3. How might multidisciplinary collaborative teams better instruct each other on their shared data? Is it appropriate in such a multidisciplinary environment that one person holds primary responsibility for all that is published in a given manuscript? Vice versa, should every author be expected to be fully cognizant and responsible for all of the data in a manuscript, even those data outside their realm of expertise?

4. ORI regulations require the institution to obtain research records and evidence at the outset of an inquiry. Should legal counsel typically be asked to assist in the data sequestration process?

5. Is it really possible to conduct a thorough assessment of the allegations where the complainant is unwilling to be involved in the process and the inquiry committee thus only hears one side of the story?

6. Could the institution have done more to try to get the complainant to become more involved? The institution could, for example, have

reminded the complainant of any applicable antiretaliation policies. If a complainant continues to refuse to come forward, should the allegation be given less weight? Why or why not?

## REFERENCES

Disis, M.L., Slattery, J.T., 2010. The road we must take: multidisciplinary team science. Sci Transl Med 2 (22), 22cm9.
Editors' Update, 2013. Research Misconduct: Three Editors Share Their Stories. Elsevier. Issue 40: http://editorsupdate.elsevier.com/issue-40-september-2013/research-misconduct-three-editors-share-their-stories/.
Gewin, V., 2012. Research: uncovering misconduct. Nature 485, 137–139.
Grens, K., 2013. What to do about "Clare Francis"? The Sci. http://www.the-scientist.com/. articles.view/articleNo/37482/title/What-to-Do-About–Clare-Francis-/.
Oransky, I., 2013. Founder of Blog Retraction Watch. http://retractionwatch.com/.
Porac, J.F., Wade, J.B., Fischer, H.M., Brown, J., Kanfer, A., Bowker, G., 2004. Human capital heterogeneity, collaborative relationships, and publication patterns in a multidisciplinary scientific alliance: a comparative case study of two scientific teams. Res Policy 33, 661–678.
Swazey, J.P., Anderson, M., Louis, K., 1993. Ethical problems in academic research. Am Sci, Article/200454131229_307_1816.HTML.

## CHAPTER 12

# Management of Research Integrity within Academic Medical Centers: A Summary and Suggested "Best Practices"

The word "integrity" appears in the title of this book, and on numerous occasions throughout this volume. Why? The word connotes a certain "wholeness" or completeness of a process, an expectation of honesty and objectivity, a code of conduct. There is no doubt that public expectations of scientists are that our practice is systematic, objective, repeatable, and exacting in process. After all, we are engaged in a process understanding the workings of nature and of our physical world, and our understanding will be distorted at best and erroneous at worst unless we practice with utmost honesty and *integrity*.

However, in recent years, there has been no escaping the clear change in the public perception of science (Macrina, 2005). The number of news stories detailing falsified scientific results, hoaxes, and misconduct investigations is clearly on the rise and made more evident both by media interest and the rapid dissemination across social media (Cornelis, 2014). Not only are these perceptions widely held publicly, but they have also begun to influence students and young scientists as they start to establish their professional identity. It seems that the most pressing scientific questions have become sufficiently complex as to require larger teams of scientists working across disciplines, with the inherent dispersion of responsibilities such growth in teams inevitably brings (Carr, 2009). And as scientific teams and enterprises have grown, has the practice of science become untenably competitive, placing undue pressures on researchers and creating the perceived need to achieve success at any cost?

A very recently published analysis on the troubling pressures affecting both the conduct of science and the training of our next generation of researchers, authored by several notable authorities including one former director of the National Institutes of Health, describes the "damaging effects of hypercompetition" in science. In their poignant editorial, they state that

*The Management of Scientific Integrity within Academic Medical Centers*
http://dx.doi.org/10.1016/B978-0-12-405198-0.00012-6

*Publishing scientific reports, especially in the most prestigious journals, has become increasingly difficult, as competition increases and reviewers and editors demand more and more from each paper. Long appendices that contain the bulk of the experimental results have become the norm for many journals and accepted practice for most scientists. As competition for jobs and promotions increases, the inflated value given to publishing in a small number of so-called "high impact" journals has put pressure on authors to rush into print, cut corners, exaggerate their findings and overstate the significance of their work. Such publication practices, abetted by the hypercompetitive grant system and job market, are changing the atmosphere in many laboratories in disturbing ways. The recent worrisome reports of substantial numbers of research publications whose results cannot be replicated are likely symptoms of today's highly pressured environment for research. If through sloppiness, error, or exaggeration, the scientific community loses the public's trust in the integrity of its work, it cannot expect to maintain public support for science (Alberts et al., 2014, p. 5774).*

As we were preparing this concluding chapter for this volume, yet another new high-profile case of data fraud, this time within the field of stem cell biology, was described in great detail in many of the world's most widely read newspapers. In this most recent case to make international headlines, a researcher working at the Brigham and Women's Hospital in Boston, MA, was found to have twice committed data fraud (by the NIH Office of Research Integrity (ORI)). Specifically, in early April of 2014 a Japanese research institute with connections to the hospital announced that one of its stem cell biology researchers had been found to have knowingly engaged in research misconduct, requiring that two recent publications in *Nature* be retracted (Weintraub, April 02, 2014). To make matters worse, just one week later—and at the very same prestigious Harvard-affiliated teaching hospital—the Research Integrity Officer (RIO) for Harvard Medical School and leadership of the Brigham and Women's Hospital announced the retraction of a major new publication in *Circulation*, as a result of "compromised data" and an ongoing misconduct investigation (Johnson, April 09, 2014). Unfortunately, the Harvard system is not unique in this regard, and these cases are unfortunately making headline news on an all too regular basis.

It is these types of cases for which the design and adherence to institutional research misconduct policies, and the oversight of the ORI, are so critically important. Fortunately, in our personal experience, the types of cases that require full reporting to ORI and the management of a full investigation are not frequent occurrences. In institutions across the country, there are thousands of scientists who are dedicated to their pursuit of science as ethical professionals, and only rarely do we find an individual who lapses in moral and ethical judgment. In fact, most of the cases described in this volume reflect management of issues that do not rise to the level of reporting to ORI, or that do not neatly reflect the narrow definition of misconduct as defined in NIH policy (see Appendix 1). It these often complex but highly nuanced matters of sometimes irresponsible, sloppy, even misguided scientific conduct that verge toward breaches of integrity that we intend as the focus of this volume for it is in their often many shades of gray that these cases have much to teach us and our students.

## 1. THEMES ILLUSTRATED IN THE CASE MATERIAL

The nine cases reviewed in this volume present complex scientific, professional, and personal circumstances that underlie the alleged scientific misconduct. These circumstances range from conflicting cultural expectations, differing levels of scientific experience, personal stresses and medical impairment, poor mentorship, and most commonly, inadequate, distorted, or failed communication. These factors, single or in combination, can be viewed as either mitigating or exacerbating factors in what, on the surface, may appear to be garden-variety scientific misconduct allegations.

The first case presents an allegation of potential data falsification/fabrication arising both out of a young investigator's failure to replicate the experiments of another scientist, a relative inexperience in understanding differing standards of methodology across fields, and a breakdown in communication between senior investigators and their younger colleagues as a lab expands its reach and numbers of scientists and becomes more productive but also more diverse and less personal and cohesive. Had the more experienced scientists in the lab taken the time to hear in more detail the young scientist's concerns, it is possible that a scientific misconduct inquiry may have been avoided and at the same time, an opportunity created for all the scientists in the lab to think carefully about why their initial published report could not be replicated. The young scientist would have also had the opportunity to talk about her perceptions of proper methodology and learn more about the variations in those standards across fields.

Another breakdown in communication coupled with different cultural expectations is at the core of case two. An apparent unlawful entry into animal facilities and inappropriate and nonapproved use of animals in experiments for an investigator no longer at the institution is a serious allegation with potential repercussions for other investigators using animal facilities. That is, every breach in both animal and human research leads to careful review of regulatory procedures and oversight and often results in more constraints and rules governing the research that impacts many scientists. The nuance in case two was how cultural expectations colored perceptions of authority that in turn "overrode" the knowledge of proper, lawful use of animal facilities. The relationships among the research team, the principal investigator, coinvestigators, research staff, and students often impact team members' behaviors in ways that seem inconsonant with their cognitive understanding of the proper and responsible standards of scientific conduct.

These relationships and lines of authority are also exemplified in case three in which a well-meaning student unable to resolve what seems like unethical behavior on the part of his principal investigator brings an allegation of scientific misconduct. Similar to the first case in this volume, the student was caught in a series of confused and confusing communications colored by the context of an increasingly busy research mentor and supervisor. Wanting to do the right thing but also, similar to the second case, follow the perceived lines of authority, the student potentially compromised his own integrity in an effort to resolve what was in reality an easily managed change in protocol. Both the student and the research supervisor were impacted by the stresses of research misconduct proceedings. While they each were for the most part absolved of wrongdoing, they also learned about issues of protocol adherence and data integrity in the cauldron of an inquiry process rather than in the more constructive and generative environment of teacher–student, senior–junior investigator relationships. Case three also illustrates the multiple regulatory agencies that are often involved in these matters and may or may not work at cross purposes.

In this latter regard, case four combines both the issues of regulatory agencies and appropriate mentorship. In this example, a young colleague, early in his career, is put in a situation for which he has insufficient training and mentorship. He is unable to accomplish the work for which he is receiving federal grant support. But he receives funding from the federal grant, and he and his research mentors report his effort as totally and productively allocated to the work as described in the grant. Both he and his mentor are caught in a difficult situation when he cannot produce the work he was assigned to do. They have in essence falsified time reports and have no product as promised in the grant to show for the time they have reported on as devoted to the work. While initially directing the blame and hence the consequences to the younger colleague, the breakdown in mentorship became more clearly the issue and more clearly the fault and fracture in the system leading to a violation in regulations.

Responsibilities around mentorship appear also in case five in which a postdoctoral fellow from another country is mentored by a faculty member who helps to translate his work for publication. In the course of the manuscript submission and peer review, a number of irregularities are noted by the reviewers thus prompting the journal to alert the institution where both the postdoctoral fellow and the senior scientist were employed. Among the allegations cited by the reviewers was the seeming use of image-editing software to alter the figures in the manuscript so as to support data and conclusions. The subsequent inquiry process found serious deviations from accepted practice and raised serious concerns about the integrity of the postdoctoral fellow. But in this matter, questions were also raised about the integrity of the senior scientist for his failure to apparently detect the inaccuracies and possible falsifications in the postdoctoral fellow's work. How complicit versus sloppy was the senior colleague? Was he trying also to advance his career or only to help a junior scientist translate his work and move forward in his career? Should he have assumed authorship on the postdoctoral fellow's work and how complicit were the other authors in the irregularities ultimately discovered in the fellow's science?

Case six picks up the theme of senior scientists helping junior scientists as well as the responsible standards for authorship. In this case, a senior scientist hires two junior colleagues who bring unpublished work from their previous institution and ask their new mentor to help them edit their draft manuscript into a version suitable for submission for publication. Their new senior colleague accepts authorship for his editorial work only to find out through an allegation of plagiarism submitted after the manuscript goes under review that the former mentor claims intellectual involvement in the work and thus a right to authorship. The former mentor alleges that the other senior colleague who trustingly accepted authorship for his editorial work is actually stealing and thus plagiarizing work from his laboratory in which the junior colleague completed the work. This case illustrates both the need for careful consideration of what constitutes the basis for authorship and the definitions of plagiarism. Further, the theme of mentorship continues in the consideration of when are senior lab directors and principal investigators entitled to authorship on work emanating from their laboratories even as their role in that work becomes less central to the finished experiments and resulting manuscript. Does a senior investigator own all data emanating from his/her laboratory regardless of whether or not he or she had a hand in the experimental design, data collection, oversight, and interpretation?

Who owns data and who can write about data collected from large, collaborative, often multi-institutional research efforts is the theme of case seven. In this case, a clinical investigator involved in a multisite trial presents data without explicit permission of the investigative team. He further presents the data incorrectly and in a way that does not acknowledge the multiple investigators' involvement as well as ownership of data by the team. The case highlights issues not only of the complexity of data ownership, especially in clinical trials in which the multiple scientists and clinicians play very different but synergistic roles, but also the often unclear guidelines for who can present on data gathered by a team and how the team approves the presentation and interpretation of their presumably jointly owned data. The case also raises the issue of impairment in an investigator as a mitigating factor in seemingly irresponsible, inappropriate, or fraudulent behavior. When does the personal circumstances of the individual come into consideration in the face of potentially serious violations of responsible scientific conduct, a question also raised in the eighth case in this volume.

In case eight, a promising, talented young physician scientist falters under unexpected personal, family, and work pressures. Feeling the pressure to succeed in her science, continue her productive trajectory, and obtain a faculty job while at the same time struggling with a sick infant and unexpected financial strains in her family, she succumbs to intentional fraudulent reporting of her work. Unable to speak with her research mentor about her personal pressures, she gets herself deeper and deeper into fraudulent scientific activity in order to delude herself and others that she is keeping up with a self-imposed pace of scientific productivity. Finally, the weariness of guilt and depression brings her to an admission to her mentor. But her integrity in the eyes of her mentor and her colleagues is seriously damaged. She loses her job, all of her work is under suspicion, and her transgressions follow her to her new job as a practicing pathologist as her former institution feels the obligation to let her new institution know about her misconduct. Similar to several of the cases in this volume, this case also illustrates the often tragic consequences of perceived or real blocks in communication between junior and senior colleagues and the inability or unwillingness to seek help and services for medical and mental health needs that may seriously compromise judgment and ability to work productively. These kinds of personal and medical needs impact scientists at all levels of experience and are more often than we realize the underlying circumstances for questionable behavior to outright ethical violations and misconduct.

Finally, the ninth case in this volume stands in contrast to each of the others in that it involves the complexities and burdens of anonymous allegations. In the eight cases outlined thus far, the complainants are known to the inquiry process and hence it is possible for the investigative officers and committees to have access to both sides of the stories and to assess the motivations and biases of all parties involved. In the ninth case, an anonymous complaint with very specific details about a recently published paper in a high-impact journal is directed to a student researcher's work but impacts multiple labs and investigators. While ultimately absolved of any wrongdoing, this case illustrates both the impact any allegation, but especially an anonymous allegation, has on young and senior researchers. Because the manuscript involved in case nine also involves multiple labs with different expertise contributing to the overall manuscript, this case also raises the increasingly common question of whether or not all authors should or can speak with authority about the full content of any manuscript. Indeed, this question is an important aspect of scientific integrity in the sense of being able to represent the content of any work in which one assumes an authorial role. But how does the increasing specialization of many aspects of biomedical science redefine (or does it redefine) this important aspect of scientific integrity?

## 2. RECOMMENDATIONS AND BEST PRACTICES

Many, if not most, of the questions raised in this volume do not have a simple or single answer but require ongoing discussion and awareness. For this reason, we provide open-ended questions at the end of each case presentation, to encourage discussion and reflection on the nuances and complexities in the exemplars provided. In the best circumstances, each discussion will shed light on appropriate approaches that may inform both individual and institutional understanding and practice. Nonetheless, in this section, we provide recommendations for best practices around communication, training in responsible scientific conduct, the process of responding to allegations, addressing underlying personal, medical, and mental health needs of both respondents and complainants, and the follow-up with participants after the adjudication of any allegation.

### 2.1 Communication

Attending to communication within and across laboratories, students, junior and senior researchers, and across institutions is perhaps the most unifying theme across each of these cases. While rarely considered in addressing issues of scientific integrity, good communication may be considered a key substrate for responsible scientific conduct and one of the central "preventative measures" for reducing the possibility of inappropriate scientific conduct and even unnecessary allegations that put investigative/inquiry processes into play. Among the best practices for communication are establishing clear and open channels for discussion in the laboratory so that all members of the lab at any level of experience feel comfortable and empowered to discuss their questions as well as their concerns. In the high-pressured context of many biomedical research laboratories, establishing this kind of supportive communication environment may be challenging. Under the ever-present deadlines for grant applications, completing and defending dissertations, preparing for scientific conferences, and the submission of research manuscripts in a timely manner, it may be difficult to convey and support an atmosphere of open communication. Further, senior scientists are often involved in multiple collaborations, may be overseeing research programs or institutes, and have heavy administrative responsibilities that limit their time and especially their ability to be more regularly available in the laboratory for spontaneous discussions and drop-in consultations.

That said, students and junior scientists learn from the example set by their more senior colleagues and principal investigators. Setting a constructive atmosphere that encourages open communication requires acknowledging good work as well as overseeing the standards and best practices of the lab. It requires understanding that students and junior scientists have more questions and need more mentoring and hence more time. Indeed, one of the basic tenets of successful mentoring is to set up consistent, predictable supervision times and to allow the space for young colleagues to speak about their challenges as well as their successes (Farmer, 2005). Laboratory meetings, office hours, one-on-one meetings, and clear reporting structures especially when delegating supervision to postdoctoral fellows, laboratory managers, or junior faculty are basic tenets of communication. It may also be appropriate for individual laboratories to have their own discussions of responsible scientific conduct and how to respond to concerns and questions. Discussing criteria for authorship generally and specifically in reference to any manuscript should be an essential part of the communication content of research groups. Waiting until a manuscript is in preparation to assign/decide upon authorship may inadvertently lead to more competitiveness and confusion among research groups.

There are multiple ways also to communicate, from face-to-face conversations and meetings to written notes and email. While more time-consuming, in-person meetings and conversations are nearly always more effective both in facilitating a supportive atmosphere for raising questions and concerns but also for addressing difficult issues such as authorship, failed experiments, data ownership, and sharing. Email and text messaging are rarely if ever effective for nuanced conversations around scientific issues. Neither allows for accurate reading or clarification of intentions and feelings in any communication. Rather, more often, email and text messages may be at the root of misunderstandings and miscommunications that are difficult to repair and more likely to escalate. On the other hand, email may be useful for summarizing procedures, discussions in laboratory meetings, and agendas for discussions. This use of email provides scaffolding for effective and open verbal communication.

## 2.2 Training in Responsible Scientific Conduct

Most academic institutions have required training for graduate students, postdoctoral fellows, and junior faculty in the responsible conduct of research. Commonly these guidelines are posted on institutional web sites. Less common is training within individual laboratories or discussions about issues of scientific integrity and scientific conduct within individual research groups. Such training in any context is critical (as well as required by federal granting agencies) and there are many excellent resources provided by ORI (http://ori.hhs.gov/general-resources-) including suggestions for courses, casebooks, guides to lab management, collaboration, authorship, and communication. These resources are good foundations for any seminar and discussion whether within an institution or a laboratory. What is important though is the tone of such training and discussions. For the training to be effective and open up communication, it should focus on the constructive practices that facilitate good science in the contemporary climate of complex collaborations, high-pressure, and intense competition. Focusing only on regulations or exemplars of unfortunate scientific misconduct does not necessarily facilitate constructive discussions and awareness. Similarly, it is also important to give examples of instances in which there were questions about scientific conduct that did not result in positive findings but rather were shaded by the factors common to many of the cases provided in this volume.

## 2.3 Responding to Allegations

We propose a series of best practices in response to allegations of scientific misconduct. Each institution will have its own procedures that are likely similar or with subtle variations that fit the culture and context of that place. We do not intend this discussion of best practices in responding to allegations to be prescriptive but rather as in the preceding and subsequent sections, to serve as springboards for discussion. We divide this section into the receipt of an allegation, the inquiry process, the investigation, considering consequences, and other claims that may warrant institutional proceedings.

### 2.3.1 Receipt of an Allegation

As illustrated by the case examples in this volume, allegations may come to the attention of an institution through many routes including through email, letter or telephone call, anonymously or by an identified complainant, from a journal editor, granting agency, supervising scientists/mentors, or collaborating scientists from other institutions. Regardless of the route of submission, the first step in addressing any allegation is to determine if it has been submitted in good faith without evident bias or personal motive. As discussed in case nine, this is especially difficult to determine with anonymous allegations, and special attention is often required before deciding to proceed further with an anonymous complaint. It is neither true that all anonymous complaints deserve attention nor that all require a response; these are case-by-case decisions.

Many institutions have an initial review step using at least two experts in the relevant field or issues to advise the RIO. While not all institutions adopt this early stage of reviewing an allegation by involving experts in the area, this procedural step does provide some check and balance against counter allegations of bias and selective filtering of allegations on the part of the RIO or senior administrative officials of the institution. These advisors have access only to the allegation as received and any relevant materials such as a copy of the manuscript referred to if applicable to the allegation as received. The task of the senior advisors is to address the basic question: "Would these allegations, if true, constitute academic misconduct and hence require further inquiry?" The advisors are not asked to comment on whether or not the allegations appear to be true but only to address the hypothetical question: "Were the allegations true?" The role of such individuals may also be to comment on whether or not the allegation is made in good faith and without evident bias or other motive. The latter is, of course, often difficult to discern but examples of questionable motive may be a colleague excluded from authorship now raising an allegation of academic misconduct. This is not to say in a situation such as this there might not be cause for concern but rather that such circumstances do bring into question the motive and good faith of the complainant.

At this stage in the process, the respondents are not made aware of the allegation. It is very possible that the allegation may not go past this stage and thus not notifying a respondent at this stage offers some protection against undue stress of knowing about any allegation toward one's work and laboratory. If the advisors answer the hypothetical question posed to them in the affirmative, the RIO and/or senior administrative officials move to the inquiry stage in the process. The senior advisors may also be helpful in advising the RIO about what data sources and other materials are important to secure and protect, at the same time respondents are notified of the allegations.

## 2.3.2 Inquiry
### Securing Records and Data

Importantly, before the inquiry process begins, the RIO in collaboration with senior administrative officials and/or legal counsel needs to take steps to secure and protect all relevant research records, databases, computers, lab samples, and other relevant evidence. There are several approaches to such a sequestration process. For example, the RIO may be clear about, sometimes with the help of the senior advisors, what needs to be secured from the complainant(s) and accomplishes this task before convening an inquiry committee. Alternatively, the RIO may wish to consult with the inquiry committee about what data and records the committee members feel are most critical and in need of protection. In that instance, the committee is convened prior to notification of the respondents and advises the RIO on the sequestration before beginning any interviews with either the respondents or complainants. In rare instances, the allegation is sufficiently serious and the risk of the respondent knowing about the allegation sufficiently imminent that the RIO may secure records and data prior to reviewing the allegation with any other advisors and/or inquiry committee.

The process of sequestration is especially delicate for all concerned. By its very purpose, the sequestration process must, in essence, be a surprise for the respondent cannot have prior notification or knowledge of the allegation lest he/she is then vulnerable to accusations of data tampering. That said, the psychological impact of meeting with an RIO and usually an attorney, learning of an allegation for the first time, and then having to produce all relevant materials, including sometimes materials and computers kept at an individual's home, may be especially stressful, anger-provoking, confusing, even humiliating for a respondent and his/her co-respondents. The RIO wants to be careful to maintain a collaborative relationship with the respondents and not create an adversarial atmosphere or perception of presumed guilt even before the inquiry process begins. It is very important that respondents understand that sequestration is as much for their protection as it is to protect the integrity of the inquiry process. RIOs need to think through the steps of the process and be prepared to spend considerable time walking a respondent through the allegation and the subsequent procedures before taking any steps to secure the relevant records and data. RIOs may also want to make themselves available for further questions, even meetings, with the respondents after the sequestration as the full impact of the news delivered in this manner begins to be fully appreciated by the individuals (often only after multiple repetitions and time to process). Commonly, respondents being notified in this manner will want to explain their actions or begin to address why they believe the allegation does not (or, more rarely, does) have merit. The RIO should be careful to stress that he/she is not the deciding individual and that such explications while understandable are best presented to the inquiry committee. Usually at the time of sequestration, all respondents are provided with a letter detailing the allegation and the membership of the inquiry committee that will be meeting with them.

Importantly, sequestration may require securing relevant e-mails and/or mirroring electronic files and computer hard drives. These computers may be used for producing manuscripts or dedicated to specific experiments in the laboratory. IT specialists may be "on call" during the sequestration process to join the RIO and legal counsel and secure the computer hard drives and other electronic media so as to make mirrored copies of each. Rarely is this possible on-site so that respondents may be without their computer overnight or for a day, and some experiments in the lab may be on hold until the experimental computer is returned. While it is important for the RIO to minimize as much as possible these disruptions on individuals and laboratories, some delay in work and unexpected inconvenience to the individual is inevitable given the seriousness and necessity of the sequestration process. In many institutions, the process of mirroring or copying all relevant electronic media does not mean the inquiry committee has access to those copies. Rather the committee must then ask the respondent to provide the relevant electronic files and e-mails. The mirrored copies are kept secure and are opened with the permission of the respondent and a senior administrative official of the institution in the event the committee suspects either subsequent data tampering on the part of a respondent or wants to search for potentially relevant files not provided by a respondent.

In this way, the potentially personal files on any respondent's computer are kept secure and not available to the process.

On the other hand, in other institutions, computers and other electronic media are considered the property of the institution and thus, a respondent's permission is not required for access. (This is not the case with a personal computer that a respondent has used for their work. Respondents may refuse mirroring of their personal computers.) In short, the approach to how sequestered material is handled in the subsequent inquiry or investigation process will be driven by the institution's policies with respect to computer use and ownership and how employees are informed about those policies. Legal counsel should always be engaged in these kinds of matters relating to the availability of sequestered materials for the inquiry and investigative processes.

Sequestration processes also often pose challenges to the confidentiality necessarily demanded by scientific misconduct proceedings. Investigators in biomedical laboratories are often housed in neighboring labs; the RIO may be well known to many investigators in his/her role; and students, postdoctoral fellows, and faculty will speculate as to why the RIO is searching a laboratory. And when lab computers are absent for even a few hours, others not involved in the process will inevitably ask questions. It is often more feasible especially in terms of protecting confidentiality to accomplish a sequestration later in the day when fewer people are around an office or laboratory facility. And whenever possible, the copying of computer hard drives and other electronic media should be accomplished overnight and returned early in the day again while fewer persons are around. Moreover, all personnel involved in the process of sequestering information (the RIO, legal counsel, IT personnel, etc.) should clearly understand the importance of preserving confidentiality of the information and the process. It is never possible in a sequestration process to prevent every potential threat to confidentiality, but holding confidentiality as a priority ensures thoughtfulness about the timing of the necessary steps.

## Notification of Respondents

If no sequestration is required, respondents should be notified by letter at the time the RIO and/or senior administrative official in the institution make the decision to move forward to an inquiry process. In some instances depending on the circumstances, the RIO may wish to hand-deliver the notification of the allegation to a respondent so as to permit some discussion at the time of the next procedural steps. In the letter notifying respondents of the inquiry process, including the letter delivered with a sequestration process, respondents are given the opportunity to identify any conflicts of interests with named members of the inquiry committee and to request a replacement of a committee member. Honoring such a request is usually at the discretion of the senior administrative official of the institution and will depend on the reasons for the respondent's concerns.

Parenthetically, many institutions have standing committees to address these matters and do not convene special committees for each allegation. The latter may allow a RIO to tailor the expertise of a committee to an allegation while the standing committee may allow more experience to emerge among committee members in the overall inquiry process. At the same time, specially appointed inquiry committees do not obligate individuals to standing time commitments and may more clearly convey by their special nature, the gravity of the situation. With standing committees, it is usually possible to seek out added consultants with special expertise as required by the nature of the science involved in the allegation.

### The Inquiry Process

The inquiry process involves more detailed information gathering. The critical question for the inquiry committee is to evaluate the likelihood that the allegations may be true and hence warrant more thorough and formal investigation. Note again that the inquiry committee is not asked to determine whether or not the allegations are true but rather that there is sufficient concern and likelihood to warrant moving forward. Hence, an inquiry committee may not need to interview every individual who can possibly shed light on aspects of the allegation or review every possible bit of evidence and data. They only need to review a sufficient amount and interview a sufficient number of persons to feel sound in their conclusion to go forward or not. Minimally, these standards will require them to interview the complainant (with some exceptions as described in the next paragraph) and the respondent(s). They may also interview those relevant to specific data collection or experiments in question who are not named as respondents. In an inquiry process, interviews are typically not recorded but committee members take detailed notes and summarize the interview, the data reviewed, and their conclusions in a report that is available to the respondents for review and comment.

It is the case that complainants may come forward with their concern but wish to remain anonymous during the inquiry process because of their relationship with one or more of the respondents or generally because of their concern about the repercussions of being perceived as a whistle-blower.

Once a complainant has come forward, the institution may assume the role of the complainant and hence protect the identity of the individual. In this instance, the complainant is not required to meet with the inquiry committee though it does preclude the committee hearing more fully the multiple perspectives on any matter.

Once the inquiry committee has completed its report (which is usually reviewed by institutional counsel), presented its conclusions and recommendations to the RIO and senior administrative official for the institution, the respondents should be given the opportunity to review and comment on the report and the committee's conclusions (though not recommendations). Usually respondents are given a set amount of time to make their comments and requests for revisions/corrections. Once the respondent has commented, the committee is asked to consider the respondent(s)' comments and revise their report if needed. Once finalized, the final copy of the report is provided to the respondent and may be also provided to the complainant depending on the circumstances. More commonly, the RIO and/ or senior administrative official for the institution notifies the complainant(s) in a letter summarizing the committee's findings and conclusions.

If the committee has not recommended going forward to investigation, there may still be recommendations for action including a letter in an employee or faculty file, a more detailed supervisory plan, or periodic mentoring meetings for a student or junior faculty. In many instances, the RIO oversees these actions, but it is also reasonable for a senior scientist in the same or closely collaborative department to assist, especially when close mentoring and supervision is recommended.

If the inquiry committee makes the recommendation to move forward to an investigation and the institution accepts that recommendation, a similar notification process for complainants and respondents occurs. It is also at this stage that the ORI is notified, a requirement in the case of NIH funding being involved in supporting the respondents' work and an option even without NIH funding. In the majority of instances, ORI will request details of the institution's inquiry process including reviewing the inquiry committee report and will follow along with the investigation process without initiating their own independent investigation. However, on notification, ORI always has the authority to initiate an independent investigation of the matter. Additionally, even if on review, ORI does not accept jurisdiction of the matter, institutions may still proceed with their investigation. That is, commonly institutions oversee more allegations regarding matters of scientific integrity and responsible conduct of research than ORI accepts under their purview.

## Investigation

The investigation process builds on the work of the inquiry committee. Typically, institutions working through their RIO appoint a special investigation committee with members different from those who have participated in the inquiry process. Indeed, while the inquiry process committee may be a standing one at many institutions, the investigation committee may be ad hoc convened especially for this purpose. There are considerable advantages to having separate committees and a new set of individuals reviewing the evidence and individuals' statements as illustrated in case six in this volume. Their weighing of the evidence and their conclusions may be different and provide the respondents the opportunity to not only make their best case but also marshal as much evidence as they have to support their points. While usually all relevant evidence has been gathered and protected early in the process, it is appropriate at the beginning of an investigation to again ask respondents if there is any additional evidence that the committee should review to ensure that any such evidence is secured. As with the inquiry process, respondents are also notified of the members of the investigation committee and given the opportunity to raise any objections or concerns about any one or more committee members.

At this stage, all interviews with the committee are recorded and transcribed. While respondents are typically not present for all interviews, they do have the opportunity to review and comment on all transcripts. Because the investigation committee is charged with determining whether or not the allegations are true usually using the standard of preponderance of the evidence, they must interview as many individuals as they feel is necessary to reach their conclusion. Similarly, they must review all the available data records and if need be, as illustrated by case eight, request repeating of experiments or coding or raw experimental data by outside experts as in case nine.

As with the inquiry process, the investigation committee prepares a report detailing all the evidence reviewed, the testimonies and their transcripts, their conclusions, and, as a separate document, their recommendations. The senior administrative official for the institution first reviews this report and then it is made available to the respondents for their comments and corrections. After the respondents have provided their comments, the investigation committee may need to meet again to review those comments in light of their assessment of the findings and their conclusions and make the necessary revisions in their report. Institutional counsel reviews the investigation committee's report before it is considered final. ORI is also provided a copy of the final report and makes a decision then regarding whether or not to accept the institution's investigation or to proceed on their own. The investigation committee's report is held as a confidential document available to the respondent(s), senior administrative officials of the institution such as the dean and provost, the RIO, members of the investigation committee, and ORI.

## Sanctions

Institutions vary in the range of consequences for academic misconduct. Some actions are so egregious as to warrant immediate dismissal from the institution, and indeed, if an investigation committee reaches a positive finding, individual respondents often leave the institution to seek employment elsewhere. In this instance, the institution is faced with the question of whether or not to notify the respondent's new institution (see below). Other times, individuals remain at the institution but may have restrictions placed on their ability to apply for grants, continue in a particular line of research, or mentor junior faculty. The type of sanction may well depend also on the level of experience of the individual. Graduate students, postdoctoral fellows, and junior faculty early in their career may require different considerations from senior scientists for whom the instance may either be isolated or only the first to be discovered. It is important to note here that the process employed when meting out sanctions sometimes can be more involved in public institutions. These issues are discussed in detail in Chapter 2. The investigation (and inquiry) committees may make recommendations to a dean or other senior administrative official about possible sanctions the committee, based on their knowledge of the case, would recommend. These recommendations are usually conveyed to the senior administrative official on a separate page from their formal report; these are not considered part of the report that is disseminated to the respondent(s) for thoughts about sanctions conveyed by a committee reflect that committee's consultative role to a dean, provost, or other senior administrative officer.

## Protecting Confidentiality

It almost goes without saying that the procedures outlined above have confidentiality as their highest priority. These proceedings may have a dramatic impact on any individual's career regardless of their stage of experience and stature in the field. Committees need to take great care not to discuss matters outside their meetings, use email sparingly, and obviously not discuss their involvement or the case with any colleagues or family members. Further, committee members should be cautioned to keep all papers relevant to the committee's work in a locked file and not transported back and forth between home and office. Institutions should explicitly provide for confidentiality of these inquiries and investigations in their policy statements. Moreover, in some cases, using a confidentiality agreement may be helpful to ensure nondisclosure and emphasize the seriousness of the confidentiality requirement. Legal counsel can assist in preparing such an agreement. Similarly, respondents need to be cautioned not to speak with others or with one another, the latter to avoid contamination of one person's account by the opinions of others. It may be especially difficult for respondents under stress not to speak about the matter with a trusted mentor or close family member, and these requests should be handled and considered on a case-by-case basis.

Inevitably, there will be breaches in confidentiality in some cases, sometimes serious enough to raise a challenge to the integrity of the process. When this occurs, unfortunate as it is, it may be warranted to start the inquiry or investigative process over with a new committee or even consider bringing scientists from other institutions in as outside consultants to hear the matter.

Not only can breaches in confidentiality hamper the institutional investigation process, as described above, but they can also, in extreme cases, lead to legal liability under theories such as invasion of privacy. Along similar lines, institutional representatives should be careful to limit any public commentary about research misconduct investigations to neutral, factual statements. Public statements capable of generating scorn or ridicule could potentially give rise to a defamation suit. These themes are discussed in more detail in the Chapter 2.

## Retaliation

Much of what we have described and recommended in this and subsequent sections pertains to the respondents in cases of scientific misconduct. But often complainants when they come forward put themselves at some risk for retaliation. Retaliation against those who bring allegations of scientific misconduct is as serious a matter as the breach of scientific integrity itself. Federal regulations require institutions to protect complainants who bring research misconduct allegations in good faith. Moreover, institutional members who retaliate against complainants may be subject to liability for civil damages.

The perceived possibility of retaliation potentially contaminates the integrity of the very process that is intended to protect scientific integrity itself. When students and junior scientists are unwilling to come forward because of concerns about their own careers or about retaliation in subtle or not so subtle forms such as poor evaluations, not being included as an author, being taken off a project, or not being recommended for a job, the entire infrastructure of scientific integrity is threatened. Hence, protecting complainants and insuring their safety and ability to proceed with their work is an important responsibility for the RIO. If the situation is sufficiently concerning or volatile, the RIO may protect the complainant by providing anonymity for the complainant and defining the institution as the formal complainant. This step should not be undertaken lightly since, as mentioned above, not having access to a complainant limits a committee's ability to obtain as full a picture of the case as possible.

## 2.4 Personal Circumstances

As illustrated in cases seven and eight, often individuals accused of scientific misconduct may be struggling with personal, medical, and/or mental health issues that are either recognized by others or unrecognized by both the individual and colleagues but discovered in the process of the investigation. These are difficult situations for sometimes regardless of the personal circumstances, the individual nonetheless was guilty of scientific misconduct as in case eight. In other instances as in case seven, the findings may not so clearly implicate the individual of clear-cut misconduct but their behavior nonetheless compromises their integrity and reputation and skirts the edges of irresponsibility. It is vitally important that RIOs know the resources of their institution that are available to employees and students for health and mental health care and that individuals involved in academic misconduct matters are made aware of these resources. Acknowledging that such matters are stressful and can impose additional stresses on families and close colleagues should be as much a part of the inquiry and investigative process as the procedures outlined in previous paragraphs for following through on the allegations. The RIO should also consider making sure the individual respondent has either access to one or more close friends or family nearby and may also consider having available a list of private physicians and mental health providers who are available to meet as needed with the individual. Respondents may also ask about obtaining legal representation. It is important to make clear that the process of investigating an internal complaint of misconduct is not like a criminal investigation, and there is no "right" to counsel; however, the RIO should not attempt to dissuade a respondent from seeking guidance if he or she feels it is appropriate. This can be a very tricky discussion with important consequences, and it is advisable to involve the institution's legal counsel. Many proceedings also allow a respondent to bring an advisor, mentor, or friend with him or her to meetings with a committee. In most institutions, that colleague usually cannot be an attorney functioning as counsel though surely if a close friend is also an attorney, he or she may accompany the respondent in their role as friend.

A word is also in order about how to combine these kinds of discussions about personal health and needs with those very necessary ones regarding procedures. This is not naturally a part of the RIO's usual activity or expectations, and some may prefer to have another person in their office or on a committee handle these kinds of discussions. If handled as a part of the RIO's ongoing communication and interaction with a respondent, the RIO will need to be thoughtful about making a clear distinction/boundary between his/her conversations about the procedures to address the allegation and discussions about the respondent's personal circumstances and needs. Issues of confidentiality will also inevitably come forward as respondents worry about whether or not their personal needs will color (positively or negatively) the scientific misconduct proceedings. It is critical that the RIO make clear that discussions about any personal needs remain confidential and not part of the inquiry/investigative proceedings unless the individual wishes to bring these forward as part of the procedural interviews. Depending on the RIO's professional background, he/she may also wish mentorship and guidance in how to handle such discussions or with whom he/she may collaborate. Many institutions have an ombudsman office who can provide these kinds of consultations both to the RIO and to the individual as needed.

## 2.5  Follow Up after the Investigation

Rarely is this area discussed in any detail in approaches to scientific integrity and conduct issues. Follow-up issues range from notification of respondents and complainants to contacting journals and former and/or current employers. Even less often discussed is attending to the impact of allegations of scientific misconduct and the associated proceedings on the careers of especially the respondents but also, in many circumstances, the complainants.

Where an investigation results in a finding of misconduct, then notification of RIO and consideration of notification to journals and other institutions is critical. Notification of journals and other institutions is clearly a case-by-case decision on the part of the RIO, the senior administrative officers of the institution, and legal counsel. Often investigations will culminate in a recommendation that a published article be retracted in order to remove fraudulent and falsified data from the publicly available scientific record and thus, correct that record as much as possible. All too often in these instances, it is simply not possible to fully remove all references to the fraudulent record and this is why the ORI's publication of documented cases of scientific misconduct (http://ori.hhs.gov/case_summary) is an essential additional step in informing the public and other scientists. When journals and editors are also the originators of the allegation, they are necessarily informed of the results of the investigative process.

Notification of former and current employers is also sometimes necessary but clearly undertaken with care and thought for the individual's reputation and career. Such action should be taken only after consultation with institutional counsel, as, in certain circumstances, communications with a new employer may create liability for tortious interference with business relations. There are instances when the breach of integrity is so egregious that the individual's entire body of work is thrown into question as with the example of Dr Das and Dr Stapel cited in the introduction to this volume. Other instances seem contained to a specific context and the individual moves on to another institution as in case eight in this volume. But there may still remain concern that the individual is not sufficiently cognizant of his or her breach so that it seems warranted to alert a new employer about a person's prior behavior. In each instance and all variations on these decisions, it is also necessary to inform the individual of all communications with their current or prior institutions.

Procedurally, it is rarely, if ever, required to follow up on the impact of scientific misconduct proceedings on the respondents whether or not there are findings of wrongdoing. If there are positive findings with the respondent remaining at their institution, usually the consequences inherently involve steps that require either ongoing or at least periodic follow-up. For example, case three in this volume resulted in a close supervision of the respondent for several years by a senior mentor. Even when allegations are not supported in an inquiry or investigative process, the individuals involved may remain unsettled and/or angry. They may resent the time lost to the investigative process, feel estranged from the complainant (if known), and/or feel that they need to move to another laboratory or other institution. While there are no standards, it is appropriate that a mentor or an RIO make it a point to follow up with a respondent in the months after the conclusion of an inquiry process where there are no findings of wrongdoing. It may also be appropriate after a sufficient period of time has passed to involve former respondents who were cleared of any wrongdoing in inquiry or investigative processes to allow them to experience the process from the other side so to speak. A former respondent also brings a potential sensitivity and thoughtfulness to the inquiry/investigative process because of his/her experience up close.

Sometimes, it is also appropriate to follow up with complainants as in, for example, cases one and three in this volume. Especially when complainants are more junior or have put themselves at some perceived risk to come forward in effect as whistle-blowers, it is important to insure their continued protection from retaliation in any form. Even without the possibility of retaliation, those who have brought forward allegations especially against close colleagues may need support in the months following the conclusion of an inquiry process. This may include speaking with a trusted mentor or friend, meeting with the RIO, or consultation with a mental health professional as needed. Additionally, especially when allegations are not supported in the inquiry process, those bringing the concerns may need reinforcement about the importance of their actions and debriefing about the process and its overall level of thoroughness and integrity.

## 3. CONCLUSION

As biomedical scientists, our behavior and professional conduct are molded by many hard years of education; we complete rigorous training in scientific methodology; and we grow comfortable with exploring the boundaries of current understanding. Unlike many professions that require its practitioners to limit their judgment and actions to what is perceived to be factual information, scientists routinely glide along the razor's edge between factual knowledge and new discovery. As such, the self-perceived ability to remain cautious in interpreting data, to recognize the limits of our own knowledge, and our deep appreciation for the awe-inspiring complexity of our natural world and the phenomena that we study, all become part of our professional identity. Most of us are proud of these finely honed intellectual skills, and our beliefs in our own objectivity rest on our adherence to proper scientific methods and the testing of refutable, falsifiable hypotheses. And yet we are all part-and-parcel of our societies and cultures, and we are prone to the same human, moral, and ethical failings that are characteristic of our species. We are also subject to the same pressures and motivations as nonscientists that cloud our judgment and our objectivity including the wish for personal success, notoriety, financial gain, professional promotion, to win a perceived race or beat out a perceived rival. Behind these and many other motivations is an implicit comparison of ourselves as scientists to other scientists. In short, the conduct of science is largely a social pursuit like any other activity that adults are engaged in for the majority of their lives. As with any other practitioner in any other profession or trade, we want to advance our careers, to be promoted within our institutions or to be attracted by enticing career opportunities elsewhere, to provide needed resources to our students and fellows, to complete important work that will be remembered long after we are gone, to provide financially for our families, and occasionally to enjoy lovely vacations. In other words, science is practiced within a social network and within the matrix of social motivations and as such, is subject to all the opportunities and pitfalls of any other human pursuit. There is neither the lone scientist accomplishing great discoveries on his or her own nor the purely objective practitioner able to accept with equal ease failed efforts and confusing results along with exciting discoveries and successful experiments and clinical trials. Science is about patience, honesty with self as well as others, and an enormous tolerance for uncertainty and ambiguity, each highly subject to the influence of others and of social pressures. And yet the pursuit of science is inherently also about growing one's career, aspiring to reach recognition for hard work, and financial security.

It is certainly true that an important motive in a number of recent cases of data fabrication or misrepresentation has been a desire on the part of one or more persons to establish priority and to receive credit for a discovery. No doubt that a great amount of fraud in the conduct and reporting of science can be traced directly to such motives, and to the highly competitive nature of practice as a scientist. The question for all of us is a simple one. How do we reach our aspirations to succeed, and to protect the benefits of healthy competition and reward, and at the same time protect the veracity of our work, the trust of our communities, and the social benefits of our discoveries when we bring our information from the lab bench to the daily newspaper? The issues raised in this volume may raise more questions than they answer—but these questions are all central to our professional lives, our conduct, and ultimately they are crucial issues to consider for any individual who is fortunate enough to be in a position to make, or report on, important discoveries that may have the potential to change lives.

# REFERENCES

Alberts, B., Kirschner, M.W., Tilghman, S., Varmus, H., 2014. Rescuing US biomedical research from its systemic flaws. Proc Natl Acad Sci 111 (16), 5773–5777.

Carr, W., 2009. Prevailing truth: the interface between religion and science. In: Snyder, P.J., Mayes, L.C., Spencer, D.D. (Eds.), Science and the Media: Delgado's Brave Bulls and the Ethics of Scientific Disclosure. Elsevier, Amsterdam, pp. 107–121.

Cornelis, G.C., 2014. It is about time we put an end to the dehumanization of the academic world. Eur J Contracept Reprod Health Care 19, 1–4.

Farmer, B., 2005. Mentoring communication. Rev of Communication 5 (2-3), 138–144.

Johnson, C.Y., April 09, 2014. Brigham and Women's stem cell study retracted over 'compromised data'.Boston Globe.URL:http://www.boston.com/news/science/blogs-/science-in-mind/2014/04/09/study-prominent-brigham-scientists-retracted-due-compromised-data/AwNP4sKvpUVed4mOCwGdxL/blog.html (accessed 25.06.14).

Macrina, F.L., 2005. Chapter 1. Methods, Manners, and the Responsible Conduct of Research. Scientific Integrity, third ed. ASM Press, Washington. pp. 1–18.

Weintraub, K., April 02, 2014. Fraud alleged in findings on stem cells. Boston Globe. URL: http://www.bostonglobe.com/metro/2014/04/01/stem/Sh5GOIUNn8Sj6ZEgv4sk3 N/story.html?p1=ArticleTab_Article_Top (accessed 25.06.14).

# APPENDIX 1

# FEDERAL POLICY ON RESEARCH MISCONDUCT[1]

U.S. Department of Health & Human Services
Office of Science and Technology Policy
[Federal Register: December 6, 2000 (Volume 65, Number 235)]

## I. RESEARCH[2] MISCONDUCT DEFINED

*Research misconduct is defined as fabrication, falsification, or plagiarism in proposing, performing, or reviewing research, or in reporting research results.*

- *Fabrication* is making up data or results and recording or reporting them.
- *Falsification* is manipulating research materials, equipment, or processes, or *changing or omitting* data or results such that the research is not *accurately* represented in the research record.[3]
- *Plagiarism* is *the appropriation of* another person's ideas, processes, results, or words without giving appropriate *credit.*
- Research misconduct does not include *honest* error or differences of *opinion.*

## II. FINDINGS OF RESEARCH MISCONDUCT

A finding of research misconduct requires that:

- there be a significant departure from accepted *practices of the relevant research community;*
- the misconduct be committed intentionally, or knowingly, or recklessly; and
- the allegation be *proven* by a preponderance of *evidence.*

---

[1] No rights, privileges, benefits, or obligations are created or abridged by issuance of this policy alone. The creation or abridgment of rights, privileges, benefits, or obligations, if any, shall occur only upon implementation of this policy by the Federal agencies.

[2] Research, as used herein, includes all basic, applied, and demonstration research in all fields of science, engineering, and mathematics. This includes, but is not limited to, research in economics, education, linguistics, medicine, psychology, social sciences, statistics, and research involving human subjects or animals.

[3] The research record is the record of data or results that embody the facts resulting from scientific inquiry, and includes, but is not limited to, research proposals, laboratory records, both physical and electronic, progress reports, abstracts, theses, oral presentations, internal reports, and journal articles.

## III. RESPONSIBILITIES OF FEDERAL AGENCIES AND RESEARCH INSTITUTIONS[4]

Agencies and research institutions are partners who share responsibility for the research process. Federal agencies have ultimate oversight authority for Federally funded research, but research institutions bear primary responsibility for prevention and detection of research *misconduct* and for the inquiry, investigation, and adjudication of research misconduct alleged to have occurred in association with their own institution.

- *Agency Policies and Procedures.* Agency policies and procedures with regard to intramural as well as extramural programs must conform to the policy described in this document.

- *Agency Referral to Research Institution.* In most cases, agencies will rely on the researcher's home institution to make the initial response to allegations of research misconduct. Agencies will usually refer allegations of research misconduct made directly to them to the appropriate research institution. However, at any time, the Federal agency may proceed with its own inquiry or investigation. Circumstances in which agencies may elect not to defer to the research institution include, but are not limited to, the following: the agency determines the institution is not prepared to handle the allegation in a manner consistent with this policy; agency involvement is needed to protect the public interest, including public health and safety; the allegation involves an entity of sufficiently small size (or an individual) that it cannot reasonably conduct the investigation itself.

- *Multiple Phases of the Response to an Allegation of Research Misconduct.* A response to an allegation of research misconduct will usually consist of several phases, including: (1) an *inquiry*—the assessment of whether the allegation has substance and if an investigation is warranted; (2) an *investigation*—the formal development of a factual record, and the examination of that record leading to dismissal of the case or to a recommendation for a finding of research misconduct or other appropriate remedies; (3) *adjudication*—during which recommendations are reviewed and appropriate corrective actions determined.

---

[4] The term "research institutions" is defined to include all organizations using Federal funds for research, including, for example, colleges and universities, intramural Federal research laboratories, Federally funded research and development centers, national user facilities, industrial laboratories, or other research institutes. Independent researchers and small research institutions are covered by this policy.

- *Agency Follow-up to Institutional Action.* After reviewing the record of the investigation, the institution's recommendations to the institution's adjudicating official, and any corrective actions taken by the research institution, the agency will take additional oversight or investigative steps if necessary. Upon completion of its review, the agency will take appropriate administrative action in accordance with applicable laws, regulations, or policies. When the agency has made a final determination, it will notify the subject of the allegation of the outcome and inform the institution regarding its disposition of the case. The agency finding of research misconduct and agency administrative actions can be appealed pursuant to the agency's applicable procedures.
- *Separation of Phases.* Adjudication is separated organizationally from inquiry and investigation. Likewise, appeals are separated organizationally from inquiry and investigation.
- *Institutional Notification of the Agency.* Research institutions will notify the funding agency (or agencies in some cases) of an allegation of research misconduct if (1) the allegation involves Federally funded research (or an application for Federal funding) and meets the Federal definition of research misconduct given above, and (2) if the institution's inquiry into the allegation determines there is sufficient evidence to proceed to an investigation. When an investigation is complete, the research institution will forward to the agency a copy of the evidentiary record, the investigative report, recommendations made to the institution's adjudicating official, and the subject's written response to the recommendations (if any). When a research institution completes the adjudication phase, it will forward the adjudicating official's decision and notify the agency of any corrective actions taken or planned.
- *Other Reasons to Notify the Agency.* At any time during an inquiry or investigation, the institution will immediately notify the Federal agency if public health or safety is at risk; if agency resources or interests are threatened; if research activities should be suspended; if there is reasonable indication of possible violations of civil or criminal law; if Federal action is required to protect the interests of those involved in the investigation; if the research institution believes the inquiry or investigation may be made public prematurely so that appropriate steps can be taken to safeguard evidence and protect the rights of those involved; or if the research community or public should be informed.
- *When More Than One Agency is Involved.* A lead agency should be designated to coordinate responses to allegations of research misconduct

when more than one agency is involved in funding activities relevant to the allegation. Each agency may implement administrative actions in accordance with applicable laws, regulations, policies, or contractual procedures.

## IV. GUIDELINES FOR FAIR AND TIMELY PROCEDURES

The following guidelines are provided to assist agencies and research institutions in developing fair and timely procedures for responding to allegations of research misconduct. They are *designed to* provide safeguards for subjects of allegations as well as for informants. Fair and timely procedures include the following.

- *Safeguards for Informants.* Safeguards for informants give individuals the confidence that they can bring allegations of research misconduct made in good faith to the attention of appropriate authorities or serve as informants to an inquiry or an investigation without suffering *retribution.* Safeguards include protection against retaliation for informants who make good faith allegations, fair and objective procedures for the examination and resolution of allegations of research misconduct, and diligence in protecting the positions and reputations of those persons who make allegations of research misconduct in good faith.

- *Safeguards for Subjects of Allegations.* Safeguards for subjects give individuals the confidence that their rights are protected and that the mere filing of an allegation of research misconduct against them will not bring their research to a halt or be the basis for other disciplinary or adverse action absent other compelling reasons. Other safeguards include timely written notification of subjects regarding substantive allegations made against *them*; a description of all such allegations; *reasonable access to the data* and other evidence *supporting* the allegations; and the opportunity to respond to allegations, the supporting evidence, and the proposed findings of research misconduct (if any).

- *Objectivity and Expertise.* The selection of individuals to review allegations and conduct investigations, who have appropriate expertise and have no unresolved conflicts of *interests, help* ensure fairness throughout all phases of the process.

- *Timeliness.* Reasonable time limits for the conduct of the inquiry, investigation, adjudication, and appeal phases (if any), with allowances for extensions where appropriate, provide confidence that the process will be well managed.

- *Confidentiality During the Inquiry, Investigation, and Decision-Making Processes.* To the extent possible, consistent with a fair and thorough investigation and as allowed by law, knowledge about the identity of subjects and informants is limited to those who need to know. Records maintained by the agency during the course of responding to an allegation of research misconduct are exempt from *disclosure* under the Freedom of Information Act to the extent permitted by law and *regulation*.

## V. AGENCY ADMINISTRATIVE ACTIONS

- *Seriousness of the Misconduct.* In deciding what administrative actions are appropriate, the agency should consider the seriousness of the misconduct, including, but not limited to, the degree to which the misconduct was knowing, intentional, or reckless; was an isolated event or part of a pattern; or had significant impact on the research record, research subjects, other researchers, institutions, or the public welfare.
- *Possible Administrative Actions.* Administrative actions available include, but are not limited to, appropriate steps to correct the research record; letters of reprimand; the imposition of special certification or assurance requirements to ensure compliance with applicable regulations or terms of an award; suspension or termination of an active award; or suspension and debarment in accordance with applicable government-wide rules on suspension and debarment. In the event of suspension or debarment, the information is made publicly available through the List of Parties Excluded from Federal Procurement and Nonprocurement Programs maintained by the U.S. General Services Administration. With respect to administrative actions imposed upon government employees, the agencies must comply with all relevant federal personnel policies and laws.
- *In Case of Criminal or Civil Fraud Violations.* If the funding agency believes that criminal or civil fraud violations may have occurred, the agency shall promptly refer the matter to the Department of Justice, the Inspector General for the agency, or other appropriate investigative body.

## VI. ROLES OF OTHER ORGANIZATIONS

This Federal policy does not limit the authority of research institutions, or other entities, to promulgate additional research misconduct policies or guidelines or more specific ethical guidance.

# APPENDIX 2

# LIFESPAN SYSTEM-WIDE POLICY

| Lifespan System-wide Policy | Subject:<br>Lifespan Policy on Research<br>Misconduct |
|---|---|

| Issuing Department:<br>Research Administration -<br>Lifespan | Latest revision date:<br>October 15, 2012 |
|---|---|

| Original Policy Date: | Approved by: |
|---|---|
| January 14, 1997 | |
| | Peggy McGill, MA, CRA<br>Administrative Director |
| | Peter J. Snyder, Ph.D.<br>Senior Vice President for<br>Research, Chief Research<br>Officer |

## PURPOSE

To provide an appropriate policy and related procedures regarding the investigation and reporting of possible research misconduct, as defined herein, and to comply with the current federal regulatory requirements applicable to research.

## 1. INTRODUCTION

### 1.1 General Policy

In all scientific and research activities, Lifespan expects the individuals performing research to observe the highest standards of honesty and professional conduct. It is integral for the enterprise of scientific and medical research to maintain the trust and confidence of both the scientific community and the public at large in the integrity of the scientific process. Unethical behavior represents a breach of confidence among scientists and researchers. It also undermines the confidence of the public and research subjects in the reliability of science and medicine. For these reasons, Lifespan considers research misconduct to be a betrayal of

177

fundamental medical and scientific principles and shall promptly deal with all instances of possible research misconduct according to the procedures set forth in this policy.

It is the goal of Lifespan to recognize when research misconduct undermines the integrity of the scientific process and the research enterprise. This policy was developed to prevent, detect, and redress research misconduct in Lifespan research programs. This policy aims to handle allegations of research misconduct swiftly and effectively, while also providing due process and fairness for those whose conduct is questioned.

## 1.2 Scope and Application

This policy and the associated procedures apply to all research activities conducted under the auspices of Lifespan, regardless of funding source. This policy applies to any person involved in research who is paid by, under the control of, or affiliated with Lifespan, such as physicians, scientists, trainees, technicians and other staff members, students, fellows, guest researchers, volunteers or collaborators at Lifespan. In addition, these policies and procedures apply to all individuals utilizing any Lifespan Institutional Review Board ("IRB") or other Lifespan research review committee for review and monitoring of research projects, regardless of whether the individuals are employed by, under the control of, or formally affiliated with Lifespan.

This policy and associated procedures will normally be followed when a Lifespan official receives an allegation. Particular circumstances in an individual case may dictate variation from the normal procedure when it is deemed to be in the best interests of research integrity, or as needed for the operations of Lifespan and/or of any relevant Federal agency. Any change from normal procedures must ensure fair treatment to the subject of the inquiry or investigation. Any significant variation from this policy and associated procedures shall be made only in consultation with the Office of General Counsel at Lifespan.

Research misconduct and/or retaliation occurring more than six years prior to submission of the allegations will not normally be investigated, unless there is compelling reason to do so. For example, an investigation may be warranted notwithstanding the lapse of more than six years when circumstances indicate (1) that the alleged research misconduct was not reasonably discoverable at an earlier time; (2) that the respondent has continued or renewed any incident of alleged research misconduct that

occurred before the six-year limitation; or (3) that the research misconduct poses a current threat to the health and safety of patients, staff, and/
or employees.

## 2. DEFINITIONS

1. *Allegation* means any written or oral statement or other indication of
   possible research misconduct made to a Lifespan official, including to a
   member of the IRB or other research review committee.
2. *Complainant* means the individual(s) who submits an allegation and/or a
   claim of retaliation.
3. *Deciding Official* means the Lifespan official who makes final determinations on research misconduct proceedings and any responsive Lifespan
   actions. At Lifespan, the deciding official is the Executive Vice President
   for physician affairs.
4. *Good Faith Allegation* means an allegation made with the honest belief
   that research misconduct may have occurred. An allegation is not in
   good faith if it is made with reckless disregard for, or willful ignorance
   of, facts that would disprove such allegation.
5. *Inquiry* means gathering information and initial fact-finding to determine whether an allegation or apparent instance of research misconduct
   warrants an investigation.
6. *Investigation* means the formal examination and evaluation of all relevant
   facts to determine if research misconduct has occurred and, if so, to
   determine the responsible person and the seriousness of the research
   misconduct.
7. *ORI* means the Office of Research Integrity, the office within the U.S.
   Department of Health and Human Services ("DHHS") that is responsible for the Research misconduct and research integrity activities of the
   U.S. Public Health Service ("PHS").
8. *Research* means a systematic experiment, study, evaluation, demonstration, or survey designed to develop or contribute to general knowledge
   (basic research) or specific knowledge (applied research) relating broadly
   to public health by establishing, discovering, developing, elucidating, or
   confirming information about, or the underlying mechanism relating to,
   biological causes, functions, or effects, diseases, treatments, or related
   matters to be studied. Research includes the development of individual
   patient case reports.

**9.** *Research Misconduct* means fabrication, falsification, or plagiarism in proposing, performing, or reviewing research, or in reporting research results.[1] It does not include honest error or honest difference in interpretation or judgment of data, or of regulatory and ethical standards.

A finding of research misconduct made under this policy requires that: (1) there be a significant departure from accepted practices of the relevant research community; (2) the misconduct be committed intentionally, knowingly, or recklessly; and (3) the allegation be proven by a preponderance of the evidence.

Good Faith allegations that describe problematic conduct that does not rise to the level of research misconduct will be handled in accordance with Section III.C of this policy.

**10.** *Research Integrity Officer* means the Lifespan official responsible for assessing allegations and determining when such allegations warrant inquiries and for overseeing inquiries and investigations. The Senior Vice President for Research/Chief Research Officer is the research integrity officer for Lifespan.

**11.** *Respondent* means the person against whom an allegation or claim of retaliation is directed or the person whose actions are the subject of an inquiry or investigation. There can be more than one respondent in any inquiry or investigation.

**12.** *Retaliation* means any adverse action taken against an individual by Lifespan or a Lifespan employee or staff member in response to a Good Faith allegation of research misconduct made by the individual or in response to good faith cooperation with research misconduct proceedings at Lifespan.

## 3. GENERAL POLICIES AND PRINCIPLES

### 3.1 Reporting Misconduct

Individuals employed by or associated with Lifespan should, in general, report suspected research misconduct to the research integrity officer. A complainant who is not comfortable bringing his or her concerns to the research integrity officer may direct those concerns to any responsible official of Lifespan, who is then required to direct the allegation to the attention of the research integrity officer. All conversations between the

---

[1]Fabrication means making up data or results, and recording or reporting them. Falsification means manipulating research materials, equipment, or processes, or changing or omitting data or results such that the research is not accurately represented in the research record. Plagiarism means the appropriation of another person's ideas, processes, results, or words without giving appropriate credit.

complainant and the research integrity officer or other responsible official will be handled confidentially, to the extent allowed by law.

If an individual is unsure whether a suspected incident falls within the definition of research misconduct, he or she may call the research integrity officer to discuss the suspected misconduct informally. The individual must be informed, however, that if the research integrity officer determines that an investigation of research misconduct is warranted, that official must submit an allegation even if the individual chooses not to do so.

If the research integrity officer believes the circumstances described by the complainant do not meet the definition of research misconduct, the research integrity officer will explain this to the complainant and, as appropriate, may refer the complainant or allegation to other offices or officials who might be helpful in resolving the problem, as further described in Section III.C of this policy. If the complainant disagrees with the research integrity officer's opinion, the complainant may still submit an allegation and it will be duly considered. If the research integrity officer believes the circumstances described do or may constitute research misconduct, he or she will advise the complainant about how to make a formal allegation.

## 3.2 Protection of the Complainant, Respondent, and Others

The rights and reputation of all parties involved in the allegation, including the complainant, must be protected throughout these procedures. Disclosure of the identities of the affected individual(s) shall be limited, to the extent possible and except as otherwise required by law, to those who need to know such identities and/or related information in order for a thorough, competent, objective, and fair research misconduct proceeding to be conducted. In addition, confidentiality must be maintained for any records or evidence from which research subjects might be identified; except as otherwise required by law, disclosure is limited to those who need to know to carry out the research misconduct proceedings.

It is Lifespan's policy that no one shall suffer retaliation for making a good faith allegation, or for providing evidence or testimony regarding the facts and circumstances surrounding alleged research misconduct during an official inquiry or investigation. Regardless of whether Lifespan or ORI ultimately determines that research misconduct occurred, the research integrity officer will monitor the treatment of individuals involved in research misconduct proceedings and will undertake reasonable efforts to protect a complainant who made a good faith allegation and others who cooperated in good faith with inquiries and investigations, including, but not limited to, all witnesses and committee members. Upon completion of an inquiry or investigation, the deciding official will determine, after consulting with the complainant, what steps, if any, are needed to restore the position or reputation of the complainant. The research integrity officer is responsible for implementing any steps that the deciding official approves.

Research misconduct proceedings will be conducted in a manner that will ensure fair treatment to the respondent(s) in thoroughly carrying out the inquiry or investigation, and shall ensure confidentiality to the extent possible without compromising public health and safety. Respondents accused of research misconduct may consult with legal counsel, or a non-lawyer personal advisor (who is not a principal or witness in the case) to seek advice and, with notice to the research integrity officer, may bring the counsel or personal advisor to interviews or meetings on the case. However, such counsel or personal advisor may provide passive assistance only to their client during official research misconduct proceedings, and may not participate in actual examination or cross-examination of witnesses.

### 3.3 Problematic Conduct that Does Not Qualify as Research Misconduct

The research integrity officer will evaluate each allegation to see if it purports to identify actions that would or might constitute research misconduct. In cases where the substance of an allegation does not rise to the level of research misconduct but involves other problematic conduct, Lifespan and the Lifespan hospitals will take such conduct very seriously and will make a decision about how best to investigate and redress it. Examples of problematic conduct that might not rise to the level of research misconduct but are none-the-less very serious, include but are not limited to: intentional or reckless disregard of or significant and substantial departure from accepted research practices, applicable federal regulations, Lifespan policies, IRB directives on the appropriate and ethical conduct of human subjects research, or recognized research ethics; or the submission of research forms or documents required by study sponsors that contain intentional or reckless material misstatements or omissions; or falsification of academic or professional credentials.

If the research integrity officer, in consultation with the Office of the General Counsel, decides not to handle a particular instance of problematic conduct as research misconduct, the matter may be referred to the applicable IRB or other research review committee, as appropriate, or to other appropriate forums within Lifespan or the Lifespan hospitals; outside consultants may also be engaged to assist with such matters, at Lifespan's discretion. On a case-by-case basis, Lifespan also reserves the right to employ the procedures in this policy to address problematic conduct that does not qualify as research misconduct.

Past precedent in the handling of particular types of problematic conduct shall not be construed to be any form of guarantee or assurance as to the way future instances of problematic conduct will be handled.

## 3.4 Role of the IRB in Problematic Conduct Involving Research with Human Subjects

If an allegation implicates research involving human subjects, the research integrity officer must consult with the Office of General Counsel and the IRB Chair to determine whether the allegation(s) should be handled by the IRB and its representatives, should be directed into the research misconduct process, or should be handled jointly by the IRB and Lifespan. If at any point in a research misconduct proceeding, the research integrity officer determines that conduct in an allegation does not constitute research misconduct, but raises concerns about the protection of human subjects in research, then the allegation will be referred to the IRB for investigation and resolution of these matters. If, in the course of IRB duties, any IRB member becomes aware of conduct that might constitute research misconduct, the Chair of the IRB will similarly consult with the research integrity officer and with the Office of General Counsel.

If it is determined by the research integrity officer in consultation with the Office of General Counsel and the IRB Chair that an allegation should be handled primarily by the IRB, such matter shall be handled in accordance with the IRB policy regarding non-compliance in the conduct of human subjects research. Pursuant to this policy, the IRB may employ any reasonable means of pursuing the investigation and resolving the matter, and for this purpose, may call upon research staff, members of the medical staff, the Office of General Counsel, or outside consultants or attorneys, for assistance. All such persons who assist for this purpose shall have full access to the relevant research materials and medical records, as an agent of the IRB itself, and all researchers and staff members of Lifespan are expected to cooperate in any such process. At the end of the process, the IRB, in consultation with the research integrity officer and the Office of General Counsel, will determine whether a violation of policies, procedures, regulations or research ethics has occurred. The IRB will then specify appropriate corrective actions (e.g., disclosure to subjects, reconsent of subjects) and may impose sanctions (e.g., temporary or permanent suspension of research at issue or of other research activities, mandatory research skills retraining). All Lifespan staff and employees are expected to comply with such determinations.

## 4. PROCESS FOR HANDLING ALLEGATIONS OF RESEARCH MISCONDUCT

### 4.1 Summary of the Research Misconduct Process

Once an allegation has been made, and once the research integrity officer, in consultation with the Office of General Counsel, has determined that the allegation purports to identify actions that constitute research misconduct and that the allegation is sufficiently credible and specific enough so that potential evidence of research misconduct may be identified, then the following procedures will be undertaken: (1) submission of the allegation and initial inquiry; (2) when warranted, an investigation to collect data and thoroughly examine the evidence; and, (3) issuance of formal findings and appropriate disposition.

If at any time during the inquiry or investigation, information is obtained that reasonably indicates the occurrence of possible criminal violations, the research integrity officer must notify the Office of General Counsel of the specific facts within 24 h. In consultation with the Office of the General Counsel, the research integrity officer, as necessary, will then promptly inform the appropriate office of the sponsoring or funding entity; ORI, if applicable; and the appropriate law enforcement officials. If reporting to ORI is applicable, ORI must also be notified promptly if any of the following conditions exist: (1) if there is an immediate health hazard involved, including a risk to human or animal subjects; (2) if there is an immediate need to protect the interests of the complainant(s) or the respondent(s) or their co-investigators and associates; (3) if there is an immediate need to protect federal resources or interests; (4) if research activities should be suspended; (5) if it is probable that the alleged incident will be reported publicly; or, (6) if the research community or public should be informed. Additional reports shall be made as required under Lifespan's Federalwide Assurance (FWA) and applicable federal, state, and local law.

Lifespan employees and Lifespan's medical staff are required to participate in any research misconduct proceedings, including reporting allegations of research misconduct as necessary, and participating in meetings and answering questions put to them, upon reasonable notice, to facilitate investigations of research misconduct. Employees have an obligation to provide relevant evidence concerning allegations to the research integrity officer or other Lifespan officials, and all agents and representatives of Lifespan with respect to the proceedings have the right to examine research and medical records relevant to the allegations. If others subject to this policy refuse to cooperate with these procedures, Lifespan will deal with this strongly, up to and including disassociation of Lifespan from research projects; revocation of all Lifespan support and/or approval; and reporting to government authorities, as required and applicable.

## 4.2 Submission of an Allegation

After consulting with the research integrity officer, a complainant may submit an allegation to the research integrity officer, or the research integrity officer may record an allegation based on information obtained from a complainant. Upon receiving or recording an allegation that purports to implicate research misconduct and that is sufficiently credible and specific enough so that potential evidence of research misconduct may be identified, the research integrity officer will promptly select an *ad hoc* committee to conduct an inquiry (the "inquiry committee"). The research integrity officer shall take steps to ensure that individuals selected to serve on the inquiry committee do not have unresolved personal, professional, or financial conflicts of interest with the respondent, complainant, or essential witnesses.

## 4.3 Inquiry

The research integrity officer will prepare a charge for the inquiry committee that describes the allegations and any related issues identified during the allegation assessment and that states that the purpose of the inquiry is to make a preliminary evaluation of the evidence and testimony of the respondent, complainant, and key witnesses to determine whether there is sufficient evidence of possible research misconduct to warrant an investigation. The purpose of this inquiry is not to determine whether research misconduct definitely occurred or who was responsible; rather, it is to determine whether more substantial investigation is warranted. Thus, an inquiry does not require a full review of all the evidence related to the allegation. Inquiry by the committee shall begin promptly after the charge is received. The research integrity officer should notify the respondent(s) of the initiation of the inquiry, and of the names of the individuals solicited to serve on the inquiry committee. The respondent may raise objections (e.g., concerns about conflicts of interest) to the individuals on the inquiry committee in writing within seven working days of the receipt of this notification, and the research integrity officer shall consider these objections. The research integrity officer shall also notify the Department Chair, Division Chief, and/or Laboratory Director of the allegations and inquiry, as appropriate.

The inquiry committee will normally interview the complainant, the respondent, and key witnesses, as well as examine relevant research records and materials. In order to avoid any claims of alteration of data, the inquiry committee will promptly attempt to locate and secure the originals of all relevant research data and/or documents if it is ascertained that such data and/or documents may be part of the case. Supervised access to the data and/or documents should be available to the respondent. The inquiry committee may employ such outside resources (e.g., legal or consulting services) as it deems appropriate to assist in the inquiry. Witness interviews should be summarized in writing, and witnesses given the opportunity to review and correct such summaries of their own statements.

All inquiries shall be completed within 60 days of initiation unless circumstances clearly warrant a longer period. If circumstances do so warrant, the record of the inquiry shall include documentation of the reasons for exceeding the 60-day period. Notwithstanding the above, the inquiry committee should not feel compelled to use the entire 60-day period if, using fair and appropriately comprehensive methods, they can come to a conclusion more quickly about whether a more substantial investigation is required.

The individuals selected to conduct the inquiry shall make every effort to be objective, impartial, and fair. The proceedings of the inquiry will be kept confidential and will not be disclosed except as necessary to facilitate a complete and comprehensive investigation.

The inquiry committee will evaluate the evidence and testimony obtained during the inquiry. Upon conclusion of the inquiry, the inquiry committee shall prepare a written report that identifies the evidence reviewed, summarizes relevant interviews, and states the conclusions of the inquiry. An investigation is warranted if there is: (1) a reasonable basis for concluding the allegation falls within the definition of research misconduct; and (2) preliminary information-gathering and fact-finding from the inquiry indicate that the allegation may have substance. The report must include sufficiently detailed information documenting the inquiry committee's recommendation as to whether further investigation is warranted. The respondent shall be provided with a copy of the inquiry committee's report and shall have 10 days to comment on it. The complainant may be notified and may be provided with relevant portions of the inquiry committee's report for comment, which shall be received by the inquiry committee within 10 days. Any comments made by the respondent or complainant will become part of the final report of the inquiry committee. Based on the comments, the inquiry committee may revise the report as appropriate.

Within 30 days of completing the inquiry, and after consultation with the research integrity officer and Lifespan's Office of General Counsel, the inquiry committee shall transmit the final report to the deciding official, who shall determine whether to initiate an investigation based on the initial findings and whether any interim administrative action is appropriate. In either case, the deciding official will notify the research integrity officer, who will then notify the respondent of the determination and provide the respondent with a copy of the final inquiry report. The research integrity officer may, in his or her sole discretion, notify the complainant of the determination and provide the complainant with relevant portions of the final report. Any previously notified Department Chair, Division Chief, and/or Laboratory Director shall also be informed of the result of the inquiry. If it is decided that an investigation is warranted, the sponsoring agency or entity and ORI, if applicable, shall also be notified. If it is necessary to notify ORI, such notification must be done in writing before the date the investigation begins, must include a copy of the final inquiry report which includes the name and position of the respondent(s), the general nature of the allegation, and the PHS application or grant numbers implicated by the investigation. The research integrity officer may also notify publications to which results of implicated research have been submitted that an investigation has been initiated.

## 4.4 Investigation

If the inquiry committee determines that further investigation is necessary, a formal investigation will be initiated within 30 days of the completion of the inquiry. The research integrity officer shall, within the 30-day time-frame, select an *ad hoc* committee (the "investigation committee") to hear the formal charges against the respondent alleged in the previously described inquiry. The research integrity officer will take steps to ensure that individuals appointed to the investigation committee do not have unresolved personal, professional, or financial conflicts of interest with the respondent, complainant, or essential witnesses. The respondent will be informed of the proposed composition of the investigation committee and will have the opportunity to raise objection to individual appointees in writing within seven working days. The research integrity officer shall consider the objections prior to appointing the investigation committee.

The investigation committee shall fully investigate the charges set forth and recommend appropriate action. The investigation shall focus on the allegations and shall examine the factual matters of the case. The investigation committee shall take steps to obtain custody of relevant research records and evidence not already secured by the inquiry committee. The investigation will normally include review of all documentation relevant to the allegation, including, but not necessarily limited to, relevant research records, computer files, proposals, manuscripts, publications, correspondence, memoranda, and notes. The investigation committee may employ such outside resources (e.g., legal or consulting services) as it deems appropriate to assist in the investigation. Interviews of the respondent, complainant, and witnesses should be tape-recorded or transcribed.

The investigation committee's charge is to generate a report that summarizes the procedures used to conduct the investigation, all of the information considered, its conclusion as to whether there is sufficient evidence to support the allegation, and any recommended administrative or disciplinary actions to be taken against the respondent in the event the allegation is substantiated. It is within the discretion of the investigation committee to incorporate by reference any report from the inquiry committee, to the extent that the investigation committee is satisfied with any aspect(s) of the inquiry committee report as constituting a comprehensive review and resolution of the issues.

All investigations should be conducted expeditiously and completed within 120 days if possible. This includes conducting the investigation, preparing the report of findings, making the report available for comment by the respondent, and submitting the report to ORI, if applicable. If the 120-day deadline cannot be met, the investigation committee shall request an extension from the research integrity officer. If applicable, a written request for an extension and an explanation for the delay must be submitted to ORI. This request to ORI shall include an interim report on the progress to date, an estimate for the date of completion of the report, and any other necessary steps. If this request is granted, periodic progress reports may also be requested by ORI. Notwithstanding the above, the investigation committee should not feel compelled to use the entire 120-day period if, using fair and appropriately comprehensive methods, they can come to a conclusion more quickly about whether research misconduct occurred, and, if so, how serious the research misconduct was and who was responsible.

The investigation committee is expected to carry its investigation through to completion and diligently to pursue all significant issues. If, for any reason, the investigation committee decides that it is appropriate or necessary to terminate the investigation, the approval of the deciding official is required. If the deciding official approves such termination, a report of the planned termination, including the reasons for the termination, shall be made to ORI, if applicable, which may then decide to undertake its own investigation.

When the investigation committee reaches a conclusion regarding an allegation, it shall submit a preliminary report reviewing all information and its conclusion to the respondent. The preliminary report shall adequately detail the evidence that supports or refutes each allegation included in the investigation. The respondent shall also be given a copy of, or supervised access to, the evidence. The respondent will have 30 days to prepare a response to the preliminary report, which shall be considered by the investigation committee before the investigation report is finalized. The investigation committee may, in its sole discretion, provide the complainant with a copy of the preliminary report or relevant portions of the report. If applicable, the complainant may be given up to 30 days to submit a response to the preliminary report, which the investigation committee shall consider in finalizing the investigation report.

After receiving the respondent and/or complainant's comments to the preliminary report, if any, the investigation committee shall prepare and maintain a final investigation report that explains the specific allegations of research misconduct, lists and adequately substantiates its findings, describes the policies and procedures under which the investigation was conducted, describes how and from whom information was obtained, and recommends the administrative or disciplinary actions to be taken against the respondent, if any. If applicable, the report shall also describe and document the PHS support, including, for example, grant numbers, grant applications, contracts, and publications listing PHS support. The final report of the investigation committee shall be made available to the respondent, who will be provided a full and fair opportunity to respond in writing to the investigation committee within seven working days of receipt of the final report. Such comments, if any, may be made a part of the record of the investigation. The final report shall also be provided to Lifespan's Office of General Counsel for review of its legal sufficiency. Comments shall be incorporated as appropriate.

The final investigation report, the complainant's and/or respondent's comments, if any, and the investigation committee's recommended administrative or disciplinary actions, if any, shall be provided to the deciding official, who will determine based on a preponderance of the evidence whether to accept the final report, its findings, and any recommended administrative or disciplinary actions. If the deciding official's determination varies from that of the investigation committee's final report, the deciding official will explain in detail the basis for rendering a decision different from that of the investigation committee. The deciding official's determination, together with the investigation committee's final report, constitutes the final investigation report for purposes of ORI review. If applicable, ORI and/or other government authorities (e.g., the federal Office of Human Research Protections and/or state agencies) should be notified of the final outcome of the investigation, and ORI shall be provided with a copy of the final report. The final investigation report provided to ORI shall describe any pending or completed administrative and/or disciplinary actions against the respondent. The complainant may be provided with those portions of the final report that address his or her role and opinions in the investigation.

## 5. CONSEQUENCES OF INVESTIGATION

### 5.1 Administrative and/or Disciplinary Actions

If the deciding official determines that the alleged research misconduct is substantiated by the findings of the investigation committee, he or she will decide on the appropriate administrative or disciplinary actions to be taken, if any, after consultation with the research integrity officer and taking into consideration the recommendations of the investigation committee. The actions may include, but are not limited to:

- withdrawal or correction of all pending or published abstracts and papers emanating from the research where research misconduct was found;
- removal of the responsible person from the particular project, letter of reprimand, special monitoring of future work, probation, suspension, salary reduction, or initiation of steps leading to possible rank reduction or termination of employment and/or medical staff privileges;
- notification to other hospitals and sponsoring agencies with which the individual has been or is affiliated, if there is reason to believe that previous research may be characterized by research misconduct; and,
- restitution of funds as appropriate to granting agencies, Lifespan, and/or research subjects.

The research integrity officer shall notify the respondent in writing of any administrative or disciplinary actions to be taken and shall also meet with the respondent to discuss the findings and the implementation of any such administrative or disciplinary actions. Any disciplinary action relating to medical staff privileges and/or IRB or other research committee review shall be coordinated with the Department Chair, Division Chief, Laboratory Director, and/or the IRB or other research review committee, as appropriate. If indicated, medical staff discipline will be pursued through established medical staff disciplinary procedures, but the procedures in this policy are distinct from, and may be taken without recourse to, medical staff disciplinary procedures.

## 5.2  Restoration of the Respondent's Reputation

If the investigation committee's finding is that no research misconduct occurred and the inquiry or investigation has resulted in any damage to the respondent's reputation, the respondent shall meet with the research integrity officer to discuss how the respondent's record shall be cleared and what reasonable efforts will be taken to restore the respondent's reputation. Any Lifespan actions to restore the respondent's reputation must first be approved by the deciding official. The implementation of such approved actions will be the responsibility of the research integrity officer. Depending on the particular circumstances, the research integrity officer should consider notifying those individuals aware of or involved in the inquiry or the investigation of the final outcome, publicizing the final outcome in forums in which the allegation was previously publicized, or expunging all reference to the allegation from the respondent's personnel file.

## 6.  OTHER CONSIDERATIONS

## 6.1  Termination of Lifespan Employment or Resignation Prior to Completing Inquiry or Investigation

The termination of the respondent's Lifespan employment or affiliation, by resignation or otherwise, before or after an allegation has been reported, will not necessarily preclude or terminate the research misconduct procedures, due to the possible compelling interests of Lifespan, research colleagues, the IRB or other research review committee, and research subjects in resolving such allegations.

If the respondent refuses to participate in the process after resignation or otherwise, the inquiry and investigation committees will use their best efforts to reach a conclusion concerning the allegations, noting in their reports the respondent's failure to cooperate and its effect on the committee's review of all the evidence.

## 6.2 Allegations Not Made in Good Faith

If relevant, the deciding official will determine whether the complainant's allegations were made in good faith. If an allegation was not made in good faith, the deciding official will determine whether any administrative, employment and/or medical staff action should be recommended against the complainant. Use of this process for malicious motives or for personal enrichment or aggrandizement shall be dealt with firmly.

## 7. RECORD RETENTION

After completion of a case and all ensuing related actions, the research integrity officer will prepare a complete file, including the records of any inquiry and/or investigation, copies of all documents and other materials furnished to the research integrity officer or the committees, and a complete record of any appeal. The research integrity officer will keep the file in a secure manner for seven years after completion of the case to permit later assessment of the case. ORI or other authorized government personnel will be given access to the records as required by law.

## 8. CONCLUSION

The integrity of a hospital and its medical staff should never be in question. Thus, Lifespan and the medical and scientific community within it must do everything possible to prevent research fraud, unethical treatment of human subjects, or other research misconduct in science and research. This policy is meant to vindicate those interests.

## 9. PROCEDURE

If a Lifespan Employee or a Lifespan Professional Staff member has a *question concerning* the *interpretation or applicability* to a particular circumstance of any of the laws or regulations referred to in this Policy, such Lifespan Employee or Lifespan Professional Staff member should first consult with his/her supervisor(s) and if his/her supervisor(s) is unable to answer the question or provide any guidance or, if, because of the circumstances, it would be inappropriate to discuss the matter with his/her supervisor(s), then such Lifespan Employee or Lifespan Professional Staff member should contact the Lifespan Senior Vice President/Chief Quality Officer; in any case, the Lifespan Employee or Lifespan Professional Staff member may contact the Office of the General Counsel or the Corporate Compliance Officer for advice. If any Lifespan Employee or Lifespan Professional Staff member is aware of any violation or threatened or potential violation of this Policy, or *suspects* a violation of this Policy has occurred, such Lifespan Employee or Professional Staff member must refer to the Policy on Code of Conduct for instruction as to what action to take. No adverse action will be taken against any party who reports, in good faith, any violation or apparent or threatened violation.

# APPENDIX 3

# POLICIES AND PROCEDURES FOR DEALING WITH ALLEGATIONS OF ACADEMIC MISCONDUCT AT YALE UNIVERSITY

*The Scholar works within an environment that has been developed for conducting, supporting, and evaluating scholarly research in the single-minded pursuit of truth. Academic fraud... is more than error; it may take the form of falsification or fabrication of data, plagiarism, or grossly negligent data collection or analysis. It is hardly possible to exaggerate the damage that can result from such a breach of the academic commitment to truth. Academic fraud... not only shatters individual careers, but besmirches the entire cause of objective research, undermines the credibility of scholarship and rends the fragile tissues of confidence between scholar and scholar, teacher and student, the university and the public...*

*All forms of academic fraud must be condemned in the strongest possible terms. All scholars have an obligation to disclose what they believe, in good faith, to be well-founded suspicions of academic fraud. Allegations of fraud must, of course, be made with great caution; yet those who come forward with such allegations must understand that the University respects the honest exercise of their judgment. At the same time, the rights of those whose scholarship or research is questioned must also be scrupulously protected, all in accord with a process that responds to such allegations with the utmost care, diligence, sensitivity, and respect for the rights of all concerned.*

Yale University first formally addressed the issue of academic fraud in its Policy Statement on Collaborative Research published in 1982 from which the above statements are taken. The University believes that the academic community must do everything within its power to guard against academic misconduct, and as part of that responsibility should have in place detailed and well-established procedures for dealing with allegations of academic misconduct in a timely and fair manner. The procedures are necessary not only to protect the academic community from fraud and other forms of misconduct, but also to protect individuals who may be unjustly accused.[1] They do not supplant existing disciplinary procedures, but rather establish an initial process for

---

[1] Retaliation against an individual for having made an allegation of academic fraud is a violation of university policy and an offense subject to discipline. On the other hand, an individual who in bad faith brings such an allegation also will be subject to discipline.

the review and investigation of allegations, which may result in the initiation of disciplinary action if warranted, in accordance with applicable procedures.

Four general principles must govern the response when allegations of academic misconduct are brought forward. They are that (1) any well-founded accusation of scholarly misconduct made in good faith must be given serious consideration; (2) an accused person must be assumed innocent until the weight of evidence requires a conclusion to the contrary; (3) appropriate effort should be made to protect the privacy and reputation of both the complainant and the respondent, to provide for fair process, and to restore the good name of an unjustly accused individual; and (4) appropriate effort should be made to preserve the integrity of the scholarly record.

The procedures to be followed are as follows:[2]

## 1. INITIATION OF ALLEGATION

**1.** Any allegation of academic misconduct, including but not limited to falsification or fabrication of data, plagiarism,[3] or gross negligence in the conduct, proposing or reporting of research or scholarship, whether lodged from within or outside the university, that is directed against an individual who, at the time of the alleged misconduct, was a faculty or staff member or postdoctoral fellow or associate of any School of Yale University, will be directed to the Dean of the School in which the respondent holds or held appointments.[4,5,6]

---

[2] For all allegations made under this policy, the Dean will consult with the Office of the General Counsel and the Provost to determine the appropriate procedures to address the allegations. When an allegation under this policy has certain connections to federal funding, federal regulations may apply. If they do, they supersede these policies and procedures to the extent necessary to avoid any violation of the federal regulations.

[3] These policies and procedures incorporate the definitions of fabrication, falsification, and plagiarism set forth in the Office of Science and Technology's Federal Policy on Research Misconduct, http://tinyurl.com/4643ucp, as clarified by the HHS Office of Research Integrity's Policy on Plagiarism, http://tinyurl.com/4lzqabq. For allegations arising under the jurisdiction of other government agencies, those agency's definitions will supersede these to the extent inconsistent.

[4] In the case of joint appointments, the complaint should be directed to the School of primary appointment; if neither School is primary, the Provost will determine to which Dean the complaint should be directed.

[5] An allegation that a student has committed academic misconduct in the course of proposing, conducting, or reporting of research or scholarship supported by the federal government will also be addressed under these procedures, and will be directed to the Dean of the School in which the respondent is enrolled. If federally supported research is not involved, the Dean may address the matter either under these procedures or under the student disciplinary procedures of the School, in the Dean's discretion.

[6] In any proceeding brought under these policies and procedures, the President may appoint an alternative university official (in place of the Dean) to supervise the process.

2. The Dean,[7] in a timely manner but generally no longer than two weeks following receipt of an allegation, with the advice of two senior faculty members designated by him or her, will determine whether there are reasonable grounds for believing that the allegation is sufficiently credible and specific so that potential evidence of academic misconduct may be identified,[8] and, if true, would constitute academic misconduct. If it is necessary to consult with others in order to make this decision, the Dean will seek to maintain, if possible, the anonymity of all individuals involved, including those who have made the allegations, and in any event will require all those consulted to treat the matter as strictly confidential.

3. If the Dean determines that there *are not* reasonable grounds for believing that the allegation is sufficiently credible and specific so that potential evidence of academic misconduct may be identified, and, if true, would constitute academic misconduct under this procedure, the Dean may dismiss the allegation altogether or may address the allegation under other applicable university policies and procedures. The results of the Dean's evaluation and the reasons for the decision will be set forth in a written report, which will be maintained for a period of three years in a locked or secured location. The Dean will promptly inform the person(s) who made the allegations of this decision.

4. If the Dean determines that there *are* reasonable grounds for believing that the allegation is sufficiently credible and specific so that potential evidence of academic misconduct may be identified, and, if true, would constitute academic misconduct, or if the Dean is unable to reach a determination in a timely manner, the Dean will appoint an Inquiry Committee to examine the allegation.

5. The Dean may secure all materials that, in his or her discretion, are directly relevant to the allegations.

---

[7] The Dean, in his or her discretion, may delegate certain ministerial functions, including informing the accused of the inquiry, and notifying the Provost and funding, regulatory, or other agencies.

[8] With regard to whether the allegations are sufficiently credible and specific, note that the mere fact that one person cannot replicate someone else's results does not constitute evidence of academic misconduct. Nor does a disagreement over interpretation or methodology rise to the level of an allegation of academic misconduct. An allegation of nonreplicability may, however, constitute important support for a specific allegation that may warrant investigation.

## 2. INQUIRY

6. The purpose of an inquiry is to conduct an initial review of the evidence to determine whether there are reasonable grounds for believing that the allegations may be true and whether an investigation is warranted. An inquiry does not require a full review of all the evidence. The Inquiry Committee will be comprised of tenured faculty members, but in consideration of the privacy of all concerned, will be limited to the minimal number required for competent completion of the task. A faculty member will not be appointed to the Inquiry Committee if the Dean believes that individual to have a potential conflict of interest in relation to the matter at hand or the individuals involved. The Inquiry Committee will immediately secure all materials it believes directly relevant to the allegations, in addition to any materials secured by the Dean (see paragraph 5 above). The Dean will promptly inform the respondent that an allegation has been made and will be subject to an inquiry and provide to him or her, the names of members of the Inquiry Committee and sufficient information about the allegation so as to permit the respondent to prepare to answer.

7. The Dean will notify the Provost that an inquiry has been initiated and of the nature of the alleged misconduct. If warranted, in the Dean's judgment, based upon an overriding need to protect the health, safety, reputation, or financial interest of others or when required by federal regulations, the Dean will notify relevant funding, regulatory, or other agencies or entities, including any journal that has published data subject to challenge, of the initiation of the inquiry. Where federally funded research is involved, the Dean will take such interim administrative actions during the inquiry (or during any later investigation) as are appropriate, in his or her judgment, to protect federal funds and ensure that the purposes of the federal financial assistance are being carried out, as required under federal regulations.

8. The Inquiry Committee will seek to interview the complainant(s) and the respondent so as to hear first hand, the allegations as well as the response of the respondent.[9] The Inquiry Committee also may interview or otherwise request information from others within or outside the university, including persons uninvolved in but knowledgeable

---

[9] When being interviewed by the Inquiry Committee, the respondent may be accompanied by an adviser, who may be any member of the Yale Community, but not by legal counsel.

about the matter under inquiry, and may seek expert advice. All persons requested to participate in the inquiry or to provide documents are expected to cooperate. The members of the Inquiry Committee will make every effort to maintain the confidentiality of their materials and deliberations. All correspondence, minutes, and other records will be marked "confidential" and kept in a locked or secured location. In seeking information, the Inquiry Committee, insofar as it can, will require confidentiality of those with whom it must communicate. The Inquiry Committee will give due regard to preserving the privacy of the person(s) bringing the allegations; however, if the Inquiry Committee determines that confidential disclosure of the identity of the complainant to any person (including the respondent) is necessary in order for it to proceed fairly with its deliberations, the Inquiry Committee, after first informing the complainant, may make such disclosure.

9. As soon as practicable, and normally within 60 days after receiving its charges, the Inquiry Committee will prepare a written report for the Dean setting forth its conclusions and the evidentiary basis for those conclusions (including summaries of interviews). If circumstances clearly warrant extending the inquiry beyond 60 days, the report shall include documentation of the reasons for the extension. The inquiry may result in one of two outcomes:

   a. If the Inquiry Committee concludes that there are not reasonable grounds for believing that the allegation may be true or that a formal investigation is not warranted, the Dean may, at his or her discretion, inform the complainant, and to the extent appropriate, persons consulted by the Inquiry Committee, of that conclusion and the reasons for it and will provide a copy of the report to the respondent. The Dean will, for a period of three years, maintain the written report of the Inquiry Committee in a locked or secured location. The Dean will also determine what steps, if any, should be taken to restore the reputation of the respondent.

   b. If the Inquiry Committee concludes that there are reasonable grounds for believing that the allegations may be true and that a formal investigation is warranted, it will provide to the Dean along with its report a formal written statement detailing the specific charge or charges of academic misconduct that should in its judgment be the subject of an investigation. The Dean will convey the report and this statement to the respondent.

**10.** If the inquiry substantiates academic misconduct that is admitted by the respondent, the Dean will determine what actions should be taken, including any further investigation, and the sanctions to be imposed or recommended, consistent with applicable disciplinary procedures and with the approval of the Provost. If fraudulent scholarship or research has been published, the Dean will see that the facts are submitted to the appropriate journals in sufficient detail to allow for correction of the relevant public record. The Dean will notify the applicable funding, regulatory or other agencies of the outcome, as required by law or regulation.[10] The Dean will notify the respondent of his or her decision in writing.

## 3. INVESTIGATION

**11.** The Dean will initiate a formal investigation, if he deems one warranted, as soon as practicable, and generally within 30 days after receiving the report from the Inquiry Committee containing its conclusions and its statement of charges. The purpose of an investigation is to gather, review, and assess all relevant evidence available to the Committee, and to determine whether the charge or charges are substantiated by a preponderance of the evidence. To constitute academic misconduct under this policy, the alleged behavior must represent a significant departure from the accepted practices of the relevant research or scholarly community and must be shown to have been committed intentionally, knowingly, or recklessly.

**12.** The investigation will be carried out by an ad hoc Investigation Committee consisting of three or more individuals whom the Dean will appoint and which may include, in addition to or in lieu of tenured members of the faculty of the School, qualified individuals from outside the School or University.[11] The Dean will not appoint to the Investigation Committee any individuals whom he or she believes have a potential conflict of interest with respect to the matter under review or the individuals involved. The Dean will inform the respondent of the membership of the Investigation Committee, and will

---

[10] In appropriate circumstances the Dean, in consultation with the Office of the General Counsel and the Provost, may notify additional persons or entities deemed to have a legitimate interest in the proceedings, including, without limitation, the current employer of the respondent.

[11] In any case, an individual that served on the Inquiry Committee may also serve on the Investigation Committee.

consider any objection promptly raised by the respondent based on conflict on interest. However, the Dean's evaluation of any such objection shall be final. The Dean will promptly inform the Provost of the initiation of the investigation and will also report the investigation to the applicable funding, regulatory, or other agencies as may be required by law or regulation.

13. The Investigation Committee will impound any materials, including laboratory notebooks, data, and other research or scholarly materials which it believes are relevant and that have not been impounded previously by the Dean or Inquiry Committee. The Investigation Committee will make every effort to hear directly from the respondent as well as the complainant(s). It may call witnesses who may be any persons within or without the university who may be knowledgeable about the matters under investigation, may seek expert advice and may review the records of the Inquiry Committee. The investigation will follow the same procedures as described in paragraph 8 to maintain confidentiality and protect the privacy of those involved. The Investigation Committee will provide to the respondent the opportunity to inspect all documents, which it will consider, except where there is a compelling need to protect the confidentiality of a communication, in which case the Committee will inform the respondent of the substance of the document. The Investigation Committee also will provide the respondent, the opportunity to submit evidence and suggest witnesses. The Investigation Committee at its discretion may permit the respondent to be present during the hearing of testimony from other witnesses or may exclude the respondent, provided that the respondent is given an opportunity to rebut or respond to the substance of all testimony.[12]

14. The Investigation Committee will complete its work as soon as possible and will ordinarily take no longer than 120 days from the date it receives its charges. If the investigation cannot be completed within that time, the Investigation Committee will report to the Dean on the status of the investigation and he or she will notify the Provost as to an

---

[12] The respondent may be accompanied before the Investigation Committee by an adviser who may be any member of the Yale University community. If the respondent has engaged legal counsel, then counsel instead will be permitted to accompany the respondent to consult with him or her as an adviser. However, the Investigation is not a trial-type proceeding and legal counsel will not be permitted to direct questions or answers or offer argument on behalf of the respondent.

estimated timetable for completion of the investigation. The Dean will also notify and request any necessary extension of time from any funding, regulatory, or other agencies as may be required by applicable law.

## 4. REPORT OF INVESTIGATION

15. Upon completion of its investigation, the Investigation Committee will prepare a written report consisting of the following three parts:

   a. A summary of the substance of the documents, the testimony, and the other forms of evidence, which the Investigation Committee relied upon in reaching its conclusion.

   b. A statement of the Committee's findings of fact and the conclusions it has drawn from those facts.

   c. The Committee's recommendation, if any, as to what actions the Dean should undertake. The report of the Investigation Committee will be adopted upon the majority vote of the members of the Committee.

16. The Dean will permit the respondent to inspect the Committee's summary of documents, testimony, and other evidence in the report and its findings of fact and conclusions, and to indicate in writing what clarification or corrections, if any, he or she believes are appropriate. The Dean, solely at his or her discretion, may allow a similar opportunity to the complainant. The Dean, after receiving the comments on the report, may ask the Investigation Committee to supplement its report. The Dean also may request the Committee to advise him or her on steps, which should be taken to restore the reputation of the respondent if the Investigation does not substantiate the charges.

   The Dean will accept the findings of fact and conclusions of the report, as supplemented, of the Investigation Committee and will determine in light of the report, what actions, including disciplinary action, he or she will take or recommend to the Provost and President be taken or initiated in accordance with university procedures. The Dean will notify the respondent of his or her decision in writing.

17. The Dean also will notify any funding, regulatory, or other agencies as required by law or regulation of the outcome of the investigation and the action that will be taken or initiated and will submit such reports as may be required. The Dean and/or Provost may determine at their discretion to notify others determined to have a legitimate interest in the outcome of the proceedings, including, for example, and without

limitation, the current employer of the respondent, if the respondent is not at Yale; coauthors of the respondent in a manuscript subject to an allegation of academic misconduct; the complainant; the home department at Yale of the respondent; and the publisher of a manuscript that was subject to an allegation of academic misconduct.

18. In the event that the investigation concludes that the charges are not substantiated, the Dean will determine what measures reasonably can and should be taken to help restore the reputation of the respondent, and will see that they are taken. If the investigation concludes that the charges were made in bad faith, the complainant will be subject to disciplinary action, in accordance with applicable procedures.

## 5. APPEAL

19. Any person accused of academic misconduct who believes that the allegation was improperly reviewed, may appeal in writing to the Provost.

20. Any appeal shall be delivered to the Provost by the date that is 30 calendar days after the date of the Dean's notice to the respondent. Additional time may be provided by the Provost in his or her discretion, and only upon prompt application and for compelling reasons.

21. In considering such an appeal, the Provost will limit his or her review to determining whether appropriate procedures and standards were applied.

September 1996, as revised May 1, 2011.

# INDEX

Printed in the United States
By Bookmasters